MilSpec Short Stories

Carl L. Adams

MilSpec Short Stories
ISBN 978-0-9833259-3-2

Library of Congress Cataloging-in-Publication Data:
Adams, Carl L., 1948 -

The inset photo for the book cover front image is by Johncairns Photos. The book cover with title design is by the author. Some cover and interior photos were obtained from the United States National Archives and Records Administration and the Australian War Museum. All images are properties of, used by permission, and/or are licensed to, the author.

This book was self-published by the author in association with Dream Weaver Ministries, LLC, Springfield, OH, USA, www.dreamweaverministries.com

Proof-edited by Crossroads Editing
www.crossroadsediting.com

Disclaimer: All of the author's personal stories contained within this book are true stories, gleaned from the author's personal files. Some stories contain only first names of characters; those names have been changed for protecting their privacy. Stories with first name characters only, which may parallel and/or match identical events to that of others, living or deceased, in society anywhere are purely co-incidental.

Dedication

This book is dedicated to God Almighty, His Son (Christ Jesus), and The Holy Spirit. Without God, I would be totally lost. Without Christ, I would have no pathway of redemption. Without The Holy Spirit, I would have no peace within.

For:

All the GIs and veterans out there
who fight, and have fought, for freedom;
May God bless and protect you all!

*"Blessed are the peacemakers,
for they shall be called the sons of God."*
(Matthew 5:9)

"It is customary in democratic countries to explore the expenditure of armament as conflicting with the social services. There is a tendency to forget that the more important social service that a government can do for its people ... is to keep them alive and free"

Sir John Slessor,
Marshall, Royal Air Force

Other books by this author:

God's Dream Catcher
God's Dream Catcher Rising
*God*Prints*
Traveling Lighted Pathways
Wanted: Lost Souls
OH MY GOD!
DreamWeaver Storybook
Issues of Life Handbook
Foot Prints of Red, White, and Blue

www.carl-adams.com
www.godsdreamcatcher.com
www.dreamweaverministries.com

Contents

The Author 10
Prologue 14
1 - Improvise 16
2 - MilSpecs 45
3 - GenSpecs 49
4 - Body Armor 53
5 - Small Arms 63
6 - Steel Pots 82
7 - Dog Tags 93
8 - Military ID Cards 102
9 - The P-38 111
10 - The Mae West 114
11 - Jeep Works 119
12 - M&Ms 127
13 - Gatling Guns 131
14 - Rations 137
15 - Dual-Use Technology 143
16 - Tech Manuals 151
17 - God's Tech Manual 157
18 - The ILS Process 160
19 - Ending Credits 170
20 - Switchology 176
21 - Identification 184
22 - The Voice of God? 190
23 - Wild Blue Yonder 198
24 - What If? 212
25 - Epilogue 227
Addendums 230

THE AUTHOR

Carl Adams is a retired U.S. Navy Logistics Specialist with over thirty-five years of service. He has also served in the U.S. Army and the U.S. Air Force. He served aboard the Aircraft Carrier USS *Independence* as a Flight Deck Troubleshooter and also flew with the 711th Special Operations Squadron as an Aerial Gunner.

Having graduated from Pensacola State College in 1982 with a degree in Liberal Arts, he has also acquired many government training courses along his military career path. Those courses further honed his knowledge, skills, and abilities in the arts of readiness, improvising, and contingency planning. Those skills propelled him to excel in the field of weapons system maintenance technology and achieve Level III Certification (highest) in the field of Military Acquisition Logistics. Achievements toward the end of his career would earn him the title of the 'Foremost Technology Transfer Specialist within the Department of Defense.' That title would be given to him by the Office of the Under Secretary of Defense for Science and Technology in 2003.

During his long career with the military, some of his direct personal services provided were Aircraft Mechanic and Electrician, Truck Driver, Infantryman, Aircraft Trades Instructor, Aircraft Examiner, Production Controller, Equipment Specialist, Aerial Gunner, Training Specialist, Logistics Specialist, and a few other specialties (not for disclosure) along the way. He has been in submarines, ships, missile silos, tanks, armored vehicles, and aircraft for deployment of military missions. He has repaired various fixed/rotary-wing military aircraft and avionics equipment, as well as having been transported in them for mission deployment. He has also been a jet backseater (Observer) and an AC-130 Spectre Gunship Aircrewman (Aerial Gunner). On the flip side of refurbishing government weaponry, he has also demolished the weapons equipment of select non-friendly agencies in sanctioned special operations tactical missions.

In the course of his military career, he sustained a traumatic head injury, nearly losing his life, while serving in the U.S. Navy. Then while serving in the U.S. Air Force, he once became lost at sea, and all but one piece of his survival equipment malfunctioned (and then also nearly losing his life). These two close call stories with death are captured in some of his books.

His last duty station was Wright-Patterson Air Force Base (WPAFB) in Dayton, Ohio, attached with the Joint Depot Maintenance Activities Group. For eleven years, his assignment was with two joint technology service units: the Joint Depot Environmental Panel (JDEP) and the Joint Technology Exchange Group (JTEG). Through JDEP and JTEG, he interfaced with many facets of the military, industry, government laboratories, the National Aeronautics and Space Administration (NASA), and academia. Their mission was seeking improved methods for cost-effective enhancements of U.S. military weapon system sustainability.

During his assignment at WPAFB, one of his collateral duties was interacting as a liaison with the acquisition community on common environmental issues which interfaced with the sustainment community of the Department of Defense (DOD). Those interfacing issues led to enormous cost savings for both of those DOD communities, and also for the establishment of a working group to explore the founding of a new organization. This new organization would address issues common to both acquisition and sustain communities within the DOD. Through his efforts, and suggestion for naming of the new organization, the Joint Acquisition Sustainment Pollution Prevention Activity (JASPPA) was established in the mid-1990s.

Since 1999, he has been engaged in jail and prison ministry. In those years, his involvement in ministry to the incarcerated has taken him into nine different correctional institutions, including Ohio Death Row. He has personally developed, and instructs, a life skills mentoring class for the incarcerated. He serves as an inmate mentor, a detention center chaplain, and is a certified chaplain with a major church denomination chaplain's commission. In addition to ministry to the incarcerated, he has worked with troubled youth since 2004 in a program called Helping Our young People connEct with God (HOPE). HOPE sessions address many of the unique problems youth face today. In each session, God will always be an added resource for resolutions to those problems discussed.

Just prior to his retirement from the U.S. Navy, he encountered an out-of-body experience, which totally reset all of his life values (this experience is also captured in some of his books). In the resetting of values, the course of his life was changed toward dedicated ministry after retirement. Thus, in his later years, he has also taken to writing books. He desires to impart the knowledge of what God has done for him to all who would read his stories. He believes a life legacy is worth leaving behind in books; values can never be placed on those inheritances. They will live on, becoming greater in value as the years pass, helping memories never to fade, and assisting others in finding their destinies.

MilSpec Short Stories is the author's tenth published book. As this book was being written and published, he began manuscripts for writing and publishing other books. He has an Author Webpage on Amazon.com, and has established a few of his own websites. Some of his Internet website addresses have previously been listed should anyone desire looking into them.

Oh, and by the way... yes, the author claims there really are aliens at WPAFB – masses work for many of them in secret!

PROLOGUE

Stories of the military intrigue most everyone, yet beyond those sacrificial battlefield tales are logistic ones. From the pages of history are revealed events which enabled soldiers and their 'war machines' to keep on fighting. From simple armor to complex weaponry, there were 'logical' processes needed being installed for development. This book explores a brief superficial look into that process, involving just a few examples of armed forces equipment using military specifications – or MilSpecs.

Within the routine operation of military activities and the lives of its members, there have arisen times when requirements for specialized equipment were suddenly needed in the field. The testing and validation cycles were drastically reduced for meeting those urgent requirements with MilSpecs. This book captures a few of those eventful chronicles for military use into the 20th century: from dog tags, to the field can opener, to M&M candy, to the GI's Jeep – any many more.

Not all of the stories in this book are concentric with the theme of MilSpecs, but they do have applications with historical specifications. *MilSpec Short Stories* was actually the beginning of the book *Foot Prints of Red, White, and Blue* (my ninth published book). However, as that book began developing, its theme took a different direction than had originally been planned. Some of its stories did not 'fit' well within the theme of *Foot Prints of Red, White, and Blue*, and so, they were withdrawn and saved for another book – which became this one.

Coffee tastes better if the latrines are dug downstream from the encampment.

(U.S. Army Field Manual of 1861)

Chapter I
Improvise

"Bulls-eye Pierre! Them boys in dat oth'r trench ov'r dare can nev'r out throw us with grenades now! We be out of deir range and be safe... hahahaha, we has a new made grenade thrower!!"

In the 1986 movie *Heartbreak Ridge*, Gunnery Sergeant Tom Highway (actor Clint Eastwood) drilled his troops in the art of improvising. Then when his troops found themselves in combat, they discovered his constant drilling of improvising paid great benefits. The 'military version' for the definition of improvising is this: the skill of making do (or getting by) with what one has available for the doing the required job until proper resources can be obtained for adequately meeting the current crisis or mission (otherwise known as being very resourceful). Improvising is actually battlefield ingenuity without engineering approval for obtaining immediate combat mission readiness.

From the earliest times of primitive combatants, it was the wisest warriors who have been resourceful in waging conflicts and winning them – even against the greatest of odds. Having a will to win and being resourceful (improvising) with what one has can rise the dawn of hope within any combatant. Those assets will foster a new horizon of destiny very difficult to defeat by any foe. In fact, that combination can wear down poorly trained foes into despair and surprised defeat. It is from that exact combination which combatants (and their associates) forged some very unusual pieces of weaponry for use in the military.

Thousands of pieces of unusual military equipment have been quickly generated for immediate mission critical needs. A few of those are mentioned briefly over the next few pages...

THE BATTERING RAM ...

Battering rams were classified as a type of siege engine – meaning a device used for overcoming city walls and then seizing the city. Battering rams originated in ancient times and were designed for breaking open the masonry walls of fortifications, and for splintering their wooden gates. Battering rams were also used in ancient Roman mines and quarries for crushing large hard rocks.

In the simplest form, a battering ram is just a large heavy log (or pole) carried by several people and used for propelling with tremendous force against an obstacle. The ramming force could be sufficient for inflicting damage to the target if the log were massive enough and/or if it were moved in rapid, successive, and impacting strikes. Some later versions of battering rams encased the logs in arrow-proof and fire-resistant canopies mounted on wheels. Inside the canopy, the log was swung from suspended chains or ropes. This improved design saved battering ram operator lives, was very effective in battle, and was also mobile.

As battering ram designs improved, they became very effective weapons of war because older wall-building materials such as stone and brick were usually weak as a result of aging. Therefore, the older walls were prone to crack when impacted with sufficient ramming force. With repeated striking blows, the cracks would grow steadily until a hole appeared in the wall. A breach would eventually become visible in the wall fabric and enable armed attackers forcing their way through the opening and engage the walled inhabitants in combat.

Battering rams had an increasingly more important impact on the evolution for the defense of city walls. Their construction would be more creative for reducing the effects of siege engines. Recorded instances for the use of battering rams with sieges of major cities include: the Destruction of Jerusalem by the Romans, the Fall of Rome, and the Siege of Constantinople. During the English Civil War from 1642 to 1651, supposedly a battering ram was used during the famous month-long siege of Gloucester in 1643. From the Siege of Gloucester came popular myths, and from its tales arose the even more famous children's rhyme, *Humpty Dumpty*.

Even though the battering ram is commonly associated with ancient times, they also have an important use in the modern era of today. Special police and military teams and armed forces often employ small one-man or two-man metal battering rams. These battering rams can be used for easily forcing open locked openings or breaching doors.

THE CATAPULT ...

A catapult is a device used for throwing or hurling an object (usually projectiles) a great distance without the aid of explosive devices. The word catapult comes from the Latin word 'catapulta' which is derived from the Greek word καταπέλτης (katapeltēs); κατα (kata) meaning downwards and πάλλω (pallō) meaning to toss or to hurl.

Catapults were thought to have been originally invented by the ancient Greeks. Their designs vary depending on the originating country of equipment development. Designated terrain of use and available materials determine the shapes and sizes of needed catapults. Even though the catapult has been used since ancient times, in various designs, it has still proven to be a most effective mechanism during warfare.

K-PATTERN FLAMETHROWER ...

The Polish K-Pattern (wzór K) Flamethrower was a mobile backpack flamethrower. They were produced in occupied Poland during WWII for the underground Polish Home Army for use in the Warsaw Uprising of 1944. Design and development on a simple flamethrower for the Polish Underground began in 1942 at the request of the Polish Home Army Main Headquarters. The design was required to be suitable for production in common workshops with readily available materials.

Its main purpose was for use against armored vehicles. Several designs were produced, but the K-Pattern was the most popular; and it became something of a 'standard weapon' with the Polish Underground. Its production was concentrated within Warsaw and, because of varied production conditions, many of these flamethrowers differed in design details. The exact numbers produced of these flamethrowers are difficult to estimate. However, reports were that about four hundred units were produced in just one workshop of Antoni Więckowski in Warsaw, Poland.

The main parts of the K-Pattern Flamethrower were two steel cylindrical tanks connected to each other; one tank for fuel and one tank for compressed air. This double-tank assembly unit was fitted with straps for making it easily transportable on the operator's back. Fuel was a mixture of diesel fuel and gasoline. A mixture valve was installed between the fuel and air tanks. The fuel gun was a pipe about two to three feet long (and was operated with a handle); was connected to the tank assembly with a fuel hose; and a dispersal valve was fitted at the muzzle.

After opening a fuel gun valve, fuel was propelled by compressed air into the fuel gun. The fuel was ignited by a simple mesh basket wrapped with a flaming rope at the muzzle of the fuel gun. The rope had to be ignited before the weapon could be used. Normal operation was short one-second bursts with only about thirty available bursts being fired. The mesh basket with the rope could be extinguished with a tin can.

Considering its primitive design and varied conditions of manufacturing, the K-Pattern Flamethrower was a very successful weapon. Its main flaw was that air pressure decreased during operation so that successive bursts were progressively shorter in effective range. The weapon could be refueled by fuel carriers and the compressed air bottle replaced with a new one in a procedure which required only about four minutes.

The K-Pattern Flamethrower is a solid testament of great human ingenuity for a needed low-cost defense weapon during a time of tremendous crisis.

The Bob Semple Tank ...

Mister Bob Semple was the Minister of Public Works for the country of New Zealand during WWII. With the Empire of Japan sweeping its threatening oppressive reign throughout the Pacific Rim in the late 1930s and into WWII, New Zealand's military needed producing tanks very quickly and from readily available materials. Working from an American postcard depicting the conversion of tractors to a tractor-tank, Mr. Semple and an engineer from the Christchurch District Works improvised the design of the tanks. Thus, the Bob

Semple Tank was conceived, designed, and built *without formal plans or blueprints*. Using resources available to Mr. Semple as Minister of Public Works, the tanks were quickly produced within the facility workshops of Christchurch.

This very simplistic tank was built from available corrugated iron on already existing tractor bases. The tank contained numerous design flaws and practical difficulties, but was never put into mass production. The concept was to predispose the tank hulls at locations ready for use in the event of a Japanese invasion; then the hulls would be mounted on tractors for use. The concept was discarded after the tanks attracted public ridicule. Despite all of these things, the Bob Semple Tank became somewhat of a symbol for New Zealand as an example of 'do it yourself' creativity for rising above obstacles when the going gets tough.

The Bob Semple Tank is an iron-clad testament of great human ingenuity for needed quick response planning and implementation during a time of tremendous crisis.

THE WZ-34 ARMORED CAR ...

The most popular armored car in the pre-WWII Polish Army was the Armored Car Type 34 (WZ-34). The Polish Army equipped its armored battalions with this car until just before the beginning of WWII, when their units were reorganized into independent armored squadrons. During the Polish Campaign, the WZ-34s were utilized some in combat; however, by then they had become obsolete, worn out, and their use was becoming limited.

The WZ-34 design was based on an earlier Armored Vehicle, the Polish Armored Halftrack Type 28. From 1934 to 1937, approximately ninety WZ-28s were converted into the WZ-34 configuration. The WZ-34 exterior was constructed of rolled and hardened sheet steel. Its interior was divided into three compartments: combat, engine, and turret. Approximately one third of them were armed with SA-37mm guns and the remaining vehicles were fitted with a 7.92mm Hotchkiss WZ-25 Machine Guns. Its major design weakness was the floor; it was constructed of wood and was not armored. The WZ-34 had a crew of two, a top speed of 31 to 34 mph, and a range of about 155 miles highway and 56 miles off road.

The WZ-34 was unique in the Polish Army in the fact that it was the only Armored Car which originally began its existence as a half-track before being redesigned as a wheeled vehicle. This redesign decision was because of two inadequate performance factors of the WZ-28 half-tracks: (1) the high level of maintenance required to keep them in the field and, (2) its low overall speed.

While fighting against the Germans in September of 1939, crews of the WZ-34s displayed tremendous bravery in spite of their vehicle limitations. The WZ-34s, which survived the Polish Campaign, were captured by the Germans and used for occupational duties after the Polish surrendered.

The Polish WZ-34 Armored Car is a steel-plated example of tremendous human ingenuity for needed quick response planning, implementation, and limited production during a time of tremendous crisis.

THE KUBUS...

The term 'Kubuś' is a Polish name for 'Little Jacob.' During WWII the name was also given to an Armored Personnel Carrier (APC) made by the Polish Home Army during the Warsaw Uprising in 1944. Its construction was only thirteen days while undergoing heavy bombing from the Germans. The vehicle's manufacturing began on 8 August 1944 in an automotive repair shop belonging to Stanisław Kwiatkowski in Warsaw. The main engineer and designer of the vehicle was Walerian Bielecki (also known as Jan). The vehicle had no prepared design specifications on paper. Instead, all vehicle construction was improvised on site, using original equipment in place with materials taken from damaged German vehicles and civilian trucks.

The vehicle chassis was fitted with steel plates for protection of the crew. The armor plates were bolted to a steel frame and then welded together. The armored car could carry eight to twelve soldiers, and was armed with a Soviet-built Degtyarev Pechotnyi (DP) 7.62 mm Machine Gun, a Polish Model K-Pattern Flamethrower, and hand grenades. Its armor was capable of stopping MG-42 and MG-34 rounds at close range.

The Kubuś armored car entered service immediately upon completion on 22 August 1944, and was attached to the Wydra Motorized Unit. The following day, together with a captured German Half-Track APC, they took part in an assault at the main campus of the Warsaw University which had been converted by the Germans into a military garrison. The assault failed but was repeated on 2 September. The Kubuś received minor damage and was successfully withdrawn from combat to Polish-held areas. However, because repairs could not be completed, on 6 September it was abandoned.

The damaged Kubuś was looted by the Germans but survived the war. In 1945, was towed to the Polish Army Museum as one of the museum's first exhibits. It was restored and is currently on display there. A full scale operational replica was created in 2004 by Juliusz Siudziński and is (as of 2009) on display at the Warsaw Uprising Museum.

The Kubuś is an iron-clad testament of great human ingenuity for a needed quick response offense during a time of tremendous crisis.

THE GUN TRUCK ...

Military gun trucks are an improvised fighting vehicle used by regular military units (or other official government armed forces). Their designs are based on conventional cargo trucks and are able to carry large amounts of weapons and armor. Their use has mainly been by regular armies for escorting military convoys within regions subject to ambush by enemy forces.

The first concept for an application with a type of gun truck was actually the Trojan Horse as mentioned in Homer's writing of the *Odyssey*. The tale was about the strategy during the Trojan War, which allowed the Greeks to finally enter the City of Troy and bring the war to an end. After the fruitless ten-year Siege of Troy, the Greeks constructed a huge wooden horse, and hid a select force of soldiers inside of it. The Greeks pretended sailing away, and the Trojans pulled the horse into their city as their victory trophy. During that night, the Greek force stealthily crept out of the horse and opened the gates for the rest of the Greek army, which had sailed back under the cover of night. The Greeks entered and destroyed the City of Troy; thus ending the long war.

The first 'modern era' gun truck was conceived and used during the Easter Rising Insurrection in 1916. The British Army used a truck fitted with an armored body; which included the smoke boxes from several steam locomotives. Cut-outs for gun access were cut in the smoke box body for allowing the troops to shoot out of the vehicle. Since the vehicle was painted black, dummy gun cut-outs were also applied for confusing snipers.

During WWII, when the German invasion of the United Kingdom seemed imminent, the British Army designed and built an improvised armored vehicle; it was called the Bedford OXA. Its design was based on the one and a half ton OXD truck, was upgraded with armor plate, and armed with a Boys .55 inch Anti-Tank Rifle and a Bren Machine Gun. Close to one thousand OXD Gun Trucks were built by 1941, and were assigned to the British Home Guard.

During the Vietnam War, the U.S. Army Transportation Corps had the responsible mission for transporting supplies from the coastal ports of Qui Nhon and Cam Ranh Bay to inland bases located at Bong Son, An Khe, Pleiku, Da Lat, and Buon Ma Thuot. The logistical requirements of the required military vehicles were enormous, often utilizing 200-truck convoys. These long convey formations became very inviting targets for Viet Cong guerrilla groups which often ambushed them in remote areas. The 8th Transportation Group was based in Qui Nhon, and often fell victim to such ambush attacks. Their supply route stretched along Route 19 between An Khe and the Mang Giang Pass, and became known to them as 'Ambush Alley.' This was because ambush incidents occurred there on an almost daily basis.

Providing security for such convoys proved virtually impossible; Military Police units did not have adequate manpower or equipment for patrolling the entire highway. In addition, other military units only patrolled the section of roads within their designated areas of operations. Thus, their own security became an additional responsibility for the transport units. At first, security was attempted with armed Jeeps; but these rapidly proved at being inadequate because of improved Viet Cong weaponry and tactics. Then on 2 September 1967, a devastating ambush killed seven drivers, wounded seventeen other GIs, and destroyed or damaged thirty trucks.

To remedy the increasingly obvious vulnerability of the supply convoys, somewhat of a 'terminator' approach for security issues was taken. A 'Hardened Convoy' concept was adopted and implemented with protection from a new type of security vehicle – The Gun Truck. This new security vehicle was based on the two-and-a-half-ton cargo truck (deuce-n-half) design. On the truck bed was two M60 Machine Guns which were protected with a barrier of sandbags. The Hardened Convoys were also reduced to only 100 trucks per convoy, with one Gun Truck for each ten transport trucks.

The Gun Truck's role in the event of ambush was to drive into the kill zone during the first few minutes of the assault; then saturate the attackers with as much firepower as possible. Early Gun Truck designs were flawed only because the sandbags became waterlogged in rains and weighed down the vehicles. Later designs replaced the sand bags with steel armor plating salvaged from scrap yards. The Gun Truck crews consisted of a driver, two gunners, a non-commissioned officer, and sometimes a grenadier armed with an M79 Grenade Launcher.

Even though transportation units had increased their own security, they still came under attacks. This forced the Gun Truck units to improve designs of their vehicles. The deuce-n-half trucks were underpowered with addition of armor and weapons slowing them down. This led to their being replaced by five-ton cargo trucks, which formed the basis for even larger Gun Trucks. The *improvised nature* of these vehicles meant they varied considerably in appearance from unit to unit. Their units gave them quite colorful nicknames such as 'Deuces Wild,' 'War Wagon', 'Iron Man,' 'Pandemonium', et cetera, which were often proudly painted on their sides in large lettering.

The armament of these larger Gun Trucks consisted of various combinations of weapons included the M60 Machine Guns, XM134 Mini-Guns, the Bofors 40mm Cannons, and .50 caliber Machine Guns. Most of these weapons were obtained by cannibalizing damaged infantry vehicles and weapons which the transport units often encountered on their return trips.

An estimated 300 to 400 trucks were 'modified' in the field in this fashion into Gun Trucks. Their use was intended as a temporary solution, but the Transportation Corps never received enough of the proposed replacement, which was the V-100 Armored Car. Therefore, Gun Trucks continued to serve until the end of the American involvement in Vietnam in 1973. With the end of the Vietnam War, the need for such vehicles disappeared and most were either scrapped or returned to their primary purpose of cargo carrying. One Gun Truck, an M54 named 'Eve of Destruction' by its crew, has been restored and is on display at the Army Transportation Museum at Fort Eustis, Virginia.

The American Gun Trucks of Vietnam paved the way for an adaption of Gun Truck design with the Hummer. Some military Hummers have gunner turrets mounted on them and the vehicles are used in a similar fashion as the Vietnam Gun Trucks were once used.

The Gun Truck is an iron-clad testament of great human ingenuity for a needed quick response offense during a time of tremendous crisis.

FLECHETTE ...

With the invention of the Gatling Gun in 1861, the military had its first grasp on a rapid-firing weapon. Before the Gatling Gun, though militaries had been disbursing various types of grapeshot from cannons (and even some shotguns) for delivering multiple rounds of shots from one single weapon. Many multiple round cannons propelled small one- to one-and-a-half-inch 'flechettes' as projectiles.

A flechette is a pointed steel projectile, which has some sort of a tail for stabilizing its trajectory. The name 'flechette' is derived from the French word *fléchette* meaning, 'little arrow' or 'dart' and in English the word is pronounced 'fleh-SHET.' Grapeshot, mini-balls, and especially flechette, were very effective in stopping dozens of soldiers with just one cannon shot.

By the time WWI occurred, the Gatling Gun design had been surpassed with the first truly automatic firing weapon – the Maxium Gun. Flechettes, though, were still very effective in combat in stopping many soldiers with one shot. When aircraft entered the military service during WWI, flechettes then began being aerially dropped over troop positions. Those little pointed steel daggers could nail holes in soldiers just like fired rifle bullets could. Flechettes were found to be a very effective combat weapon and low in production cost – a very attractive weapons bargain.

During WWII, both Allied and Axis Powers experimented with further developing the use of flechettes. Experimentation went from using fletchettes to 'bomblets' which were small iron bombs with no explosive charges. The bomblets weighed up to around twenty-five pounds and were loaded on bombers by the thousands for being dropped over cities and industrial areas. Picking up energy from high altitude drops, the small heavy bomblets could devastate entire buildings by literally ripping through anything in their path. Entire sections of cities were flattened by bomblets. However, as the war progressed, flying bomber aircraft became increasingly more dangerous because of their heavy losses. Therefore, tactical bombing was concentrated on bombs having high explosive charges because of their far-reaching lethal range with associated collateral damage. In contrast, flechettes and bomblets only destroyed what they came in contact with.

After WWII, attention was again turned to the use of flechettes. Flechettes could still be used in areas where stopping the enemy (or offenders) was the objective, while yielding minimum collateral damage to surrounding equipment. Examples of such areas could be while transporting prisoners, riot control, in office spaces, and many other restrictive places requiring certain levels of safety. Experimentation did develop some good flechettes for those applications. Those experiments, though, collaterally led to the development of what became known as 'Lazy Dog Bomblets.' They were also sometimes called 'Lawn Darts' or Buzz Bombs.' They were very small, unguided, and unpowered missiles, each measuring 1.75 inches in length, half an inch in diameter, and weighing 0.7 of an ounce.

Originally the U.S. Army Armament Laboratory Program code named the special flechette project 'Lazy Dog' and that is how the bomblet weapon first acquired its name. The project's objective was to design and test free-fall small missiles, and their dispensing units, for use in bomber and fighter aircraft. Thus, Lazy Dog Bomblets became designed for being dispersed over battlefields from bomb rack cluster adapters mounted under aircraft wings.

Technically, Lazy Dog Bomblets were actually not bombs because they contained no explosives. Yet, in many applications they proved equally as destructive. Lazy Dog projectiles of various shapes and sizes were tested at the Air Proving Ground Center at Eglin Air Force Base, Florida, from late 1951 to early 1952. An F-84 flying at 400 knots and seventy-five feet above the ground served as the experimental testing device, while an old Jeep and a B-24 aircraft were used as ground targets. Results were eight hits per square yard of surface area.

Testing revealed that two of the experimental design shapes were the most effective. Testing further revealed that when the projectiles impacted, they could penetrate nearly any material upon impact. Some reports documented their speeds *often exceeded 500 mph* before impact. One shape had the force of a .50 caliber bullet and could penetrate twenty-four inches of packed sand. The other shape could penetrate twelve inches of sand, which was twice as much as a .45 caliber bullet when fired point blank. One of the tested Lazy Dog designs was shipped to the Far East Air Force (FEAF) for combat use by mid-1952. Upon arrival, the FEAF immediately ordered 16,000 Lazy Dog Bomblet weapon systems.

In the late 1950s, America had entered involvement in the Vietnam conflict. This conflict provided further 'rationale' for testing and using Lazy Dogs in that conflict. During the Vietnam War, reports regarding flechette varied. Ground troops generally disliked shotgun fired flechette because of dense jungle foliage. Yet, they liked the aerial dropped Lazy Dogs because they were highly effective against enemy troops hidden beneath the jungle canopy. Those small bomblets could drive right through the jungles and nail the enemy fixed in their positions. In addition, those little steel munitions were very cheap and easy to scatter over large areas.

However, their effects were often quite indiscriminately gruesome.

Lazy Dog Bomblets were always dropped in very large numbers. From almost any kind of flying vehicle, they could be hurled from buckets, dropped by hand, or thrown in their small paper shipping bags. Usually, though, they were dispersed from an aircraft wing-mounted bomb rack (Mark 44 Cluster Adapter). The Mark 44 Adapter was a very simple hinged casing with built-in bins for holding the projectiles. The bins were opened by a mechanical time-delay fuse. The adapters themselves were about seventy inches long and fourteen inches in diameter. They were shipped empty and then filled upon mission demand. Depending on how many projectiles could be packed in the Mark 44, its loaded weight varied between 560 and 625 pounds. Just one packed Mark 44 is listed as having the 'theoretical' maximum number of Lazy Dog projectiles as an astonishing capacity of 17,500 bomblets!

No soldier would want to be underneath the sky with one Mark 44 opening on the way down – not to mention several of them!

Since the use of flechettes and Lazy Dog Bomblets during the Vietnam War, there has been great controversy regarding their use as to whether they have violated the Geneva Convention. This is because their use is so very silent, highly lethal, and vastly demoralizing to ground troops. One only has to imagine being in a relatively quiet combat setting and all of a sudden, surrounding leaves and tree branches explode and fall to the ground. Then many nearby troops suddenly become traumatically injured or are dead. In this scenario, the Lazy Dog Bomblet could possibly qualify as a weapon of immense terror if used on very large scales.

While the flechettes and Lazy Dog Bomblets were being experimented with during the 1950s, the U.S. Army entertained the concept of a flechette-based individual weapon within the Army's Project SALVO. One purpose of Project SALVO was to explore, design, and develop a small weapon with a high rate of fire which would be considerably deadlier than the larger weapons being developed in the 1950s. From that experimentation, the Special Purpose Individual Weapon (SPIW) was born. The SPIW was a long-running Army Program to develop (in part) a workable flechette-based rifle (with other concepts involved). Those concepts continued testing into the 1980s and 1990s under the Future Rifle Program and the Advanced Combat Rifle program. However, results from neither program yielded a system useful enough for warranting the replacement of the standard M-16 Assault Rifle.

In various forms and applications, flechettes have been around since the Chinese invented the cannon around the 6th century. Even though the use of flechettes have become 'horrific' and legally limited in warfare, it is still in use centuries later. And, it can still be as cheap as it ever was – relatively speaking (and in MilSpec terms too).

THE TOILET BOMB ...

The military has always been ingenious when it comes to getting the job done without the properly authorized equipment. Why? They have a 'can do' attitude most of the time. The entries previously mentioned for this chapter are, of course, just a minuscule scratch on the surface of all the ingenious things GIs have come up with for fighting their conflicts. This next entry, though, is one of great humorous thought from the sailors of the USS *Midway* (CVA 41).

In October of 1965, U.S. Navy Commander Clarence 'Bill' Stoddard, Jr. was the Executive Officer of VA-25 (Fixed Wing Attack Squadron Twenty-Five) and was flying A-1H Skyraider airplanes. VA-25's motto was 'Fist of the Fleet' and they boasted along with the aircraft manufacturer (Douglas Aircraft) that with the Skyraider could carry anything but the kitchen sink. VA-25 proudly lived up to that phrase with carrying a special bomb to the North Vietnamese in commemoration of the six-millionth pound of ordnance having been dropped. This particular bomb to be dropped, however, was of a very unique type – it was a toilet!

The toilet was a damaged one which was going to be thrown overboard. One of the Plane Captains rescued it and then the Ordnance Crew made a special ordnance rack for it, complete with tailfins and a nose fuse. The flight deck Aircraft Checkers maintained constant position for blocking the strange view from the Air Boss and the Ship's Captain while the aircraft was being taxied forward to the catapult for launch. Just as the aircraft was being launched off the ship, the pilot and launch crew received a message from the bridge, 'What the hell was on 572's right wing?'

The flight was a tactical sortie strike going to the Vietnam Delta. When the squadron arrived in the target area and Commander Stoddard was reading the ordnance list to the Forward Air Controller (FAC), he ended it with 'and one code named Sani-Flush'. The FAC couldn't believe it and positioned himself to see it. The special bomb was dropped in a dive with a wingman flying a tight wing position for filming the illustrious drop. When the toilet came off, it turned bowl hole into the wind and nearly struck his aircraft. The subsequent filmed event made an outstanding Pilot's Ready Room movie. The FAC reported that the commode whistled all the way downward into its impact.

In all branches of the military, there are always numerous jokes with germ warfare. However, with this special event bombing from the USS *Midway,* VA-25, and delivered by Commander Stoddard, military air intelligence took germ warfare toward sweepingly new explosive heights.

During a later mission, on 14 September 1966, while acting as Commanding Officer of VA-25 embarked on the USS *Coral Sea* (CVA 43), Commander Stoddard was leading a two-plane bombing mission over North Vietnam searching for enemy truck convoys. While in the area of Nghi Thiet, his radar detection system warned of enemy missile activity. Taking immediate evasive action, he withdrew over the Gulf of Tonkin before his plane was struck by enemy surface-to-air missiles.

Commander Stoddard was initially listed as Missing in Action (MIA), but in 1973 the military changed his status to KIA. Incidentally, plane he was shot down in was the same one he had dropped the toilet with – that famous Douglas Aircraft A-1 Skyraider, Aircraft Number 572.

Japanese model maker, Hasegawa Models, later cast a 1/72nd scale model of Commander Stoddard's A-1 Skyraider Aircraft, and also complete with a wing mounted bomb toilet included. The new model kit was called a 'Limited-Release Variation' and was sold for about $19.00.

Toilet Bomb on Commander Stoddard's A-1 Skyraider

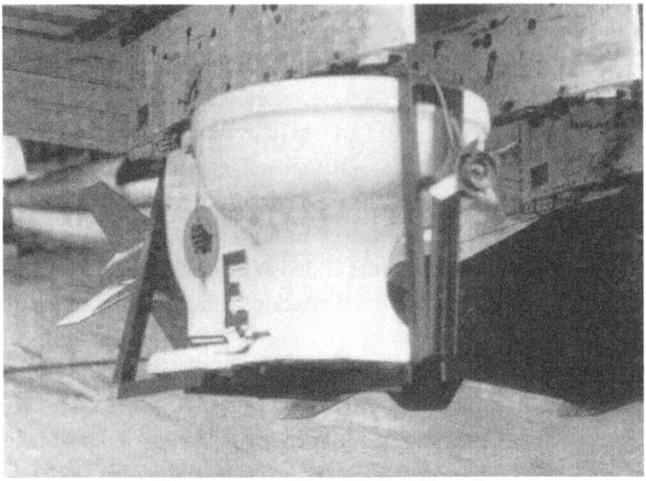

Close-up of the Toilet Bomb

Covered Battering Ram

Catapult of the Middle Ages

Polish K-Pattern Flame Thrower

New Zealand 'Bob Semple' Tank

Polish WZ-34 Armored Car

Polish Kubuś Armored Personnel Carrier

British Gun Truck used in Easter Rising Insurgency

WWII British Bedford OXA Gun Truck

WWII British Bedford OXD Gun Truck

USA Gun Truck used in Vietnam

1914 Flechettes

Lazy Dog Bomblets

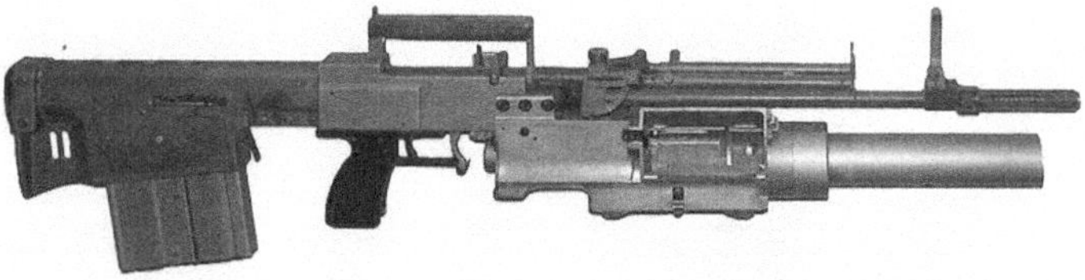

Special Purpose Individual Weapon (SPIW)

Chapter II
MilSpecs

"Guys, there really was a plan, somewhere. However, think someone somewhere got the specs mixed up"

The acronym 'MilSpec' is derived from the U.S. military term called a Military Standard (MIL-STD). Military Specifications (MIL-SPEC) are the layers of data within MIL-STDs which are used for helping achieve standardized objectives throughout the entire operational system of the military. MIL-SPEC is the proper acronym for the term in reference to military standards and the informal use is MilSpec.

Standardization is not only beneficial in achieving interoperability within all the branches of the military, but also ensures every usable product meets certain specified requirements of commonality, reliability, life cycle cost, logistics compatibility, etc. MilSpec standards are also used by other non-defense organizations in industry. According to the Government Accountability Office (GAO), MilSpecs describe the physical and operational characteristics of a product, and detail the processes and materials for making the product.

MilSpecs evolved from the need to ensure proper performance, maintainability, and reparability (MRO), and logistical usefulness of military equipment. The two goals of MRO and logistics favor certain general concepts, such as interchangeability, standardization (of equipment and processes), cataloging, communications, and training. In the late eighteenth century and throughout the nineteenth, the American and French militaries were early adopters and longtime developmental sponsors and advocates of interchangeability and standardization of military equipment. By World War Two (WWII), virtually all national militaries and alliances were users of the same systems of industrial processes. Both Allied Forces and Axis Powers were busy standardizing and cataloging the logistics of their weapon systems. Two prime Allied examples were: the AN Cataloging System (a U.S. spawned Army-Navy system), and the DEF-STAN Cataloging System (a British spawned Defence Standard System).

During WWII, American screws, bolts, and nuts did not fit British equipment properly, and therefore, were not fully interchangeable. Thus, use of MilSpecs went on to provide many benefits, such as minimizing the number of types of ammunition, ensuring compatibility of tools, and ensuring quality during production of military equipment. Such use resulted in ammunition cases which could be opened without tools.

This was a basic beginning for an effort in standardizing differences with dimensional tolerances across varied weapon systems and manufacturing bases. That beginning, however, generated the proliferation of standards bringing many disadvantages. Arguments came that the large number of standards (nearly 30,000 by 1990) imposed needless restrictions, increased contracting cost (thus military cost as well), and impeded the incorporation of latest technologies. Responding to increasing criticism, U.S. Secretary of Defense William Perry issued a memorandum in 1994 prohibiting the use of most MilSpecs without waivers, and it became known as the 'Perry Memo'.

Many MilSpecs were subsequently canceled and in their place the U.S. Department of Defense (DOD) encouraged use of industry standards. Industry Standards ISO 6000, 9000, and 12000 series, as well as SAE standards such as the AS 9100 and AMS 2404 became very common. Weapon systems were required to use 'performance specifications' that described the desired features and performance of the weapon, as opposed to how those goals would be reached. In 2005, the DOD issued a new memorandum which eliminated the requirement to obtain waivers in order to use defense standards. The 2005 memo did not reinstate any canceled MilSpecs. According to a 2003 issue of *Gateway*, published by the Human Systems Information Analysis Center, the numbers of MilSpecs were reduced from 45,500 to 28,300.

MilSpecs came into use at a crucial time in world events of industrial revolution. *MilSpecs, though, enabled both the military and industry in paving the way toward cost effective process standardization and manufacturing operations for achieving common goals.*

In coming to an end with this very brief informational dialog about MilSpecs, it could be asked why the author would provide an entry about such a topic in a book with military short stories. The concept for the principle of MilSpecs provides the basic structure for understanding a partial theme weaving through some of the framework within this book. There is also an application with MilSpecs beyond military life; they parallel the principle of a disciplined life. A life governed by a set of regimented policies which have been tried and tested, *is generally not subject to being in disarray for any length of time*.

With all the pro and con rhetoric arising from defense standards, there is one bottom line with the universal use of MilSpecs. It could be said they brought unity (in some respects) to a rising industrial world. MilSpecs were actually a critical means of bringing the world together for the beginnings of speaking commonality within everything its operations hoped to accomplish – then, and for the future.

CHAPTER III
GENSPECS

"Hey up there Leftentant, you sure that spec was the correct one for this operation? Things around here sure ain't lookin' so good!"

The term 'GENSPEC' is an acronym for General Specification. By definition it is a specification which covers requirements for two or more classes, grades, materials, services, styles, or types of goods or services. A GenSpec can be used by academia, industry, and government agencies. Each agency can apply its own specific number in addition to an overall referenced GenSpec number, thus, identifying it to their respective agency, and application.

As MilSpecs evolved from the need to ensure proper performance, maintainability and reparability (MRO), and logistical usefulness of military equipment, so did the use of GenSpecs. GenSpecs became an integral part of the logistical process for development of military equipment. GenSpecs are referenced for repair of common equipment using common tools, hardware, and general processes, where specific details need not be outlined. In some procedures, GenSpecs are integrated along with detailed MilSpec applications. This practice reduces 'reinvention' of maintenance manual outlining, and also streamlines the overall standardization of logistical processes.

In my days of aviation maintenance, I have been involved with the implementation of five different GenSpecs for repair of U.S. Navy and Marine Corps aircraft. All five GenSpecs were for reduced cost while improving the repair of the A-4 and H-3 aircraft.

Once upon a time, while performing depot level repair on the U.S. Navy and Marine Corps SH-3 Aircraft, I observed some serious chaffing of the wiring harness to the pilot's Bearing Distance Heading Indicator (BDHI). I'd thought about wrapping and tying the wiring bundle, but in doing so, noticed a hydraulic line directly underneath the chaffing wiring bundle. Knowing that hydraulic fluid can be flammable when under high pressure and under the right conditions, I began conducting some research.

Only one certain part numbered BDHI was issued from the supply system with the longer canister housing; and it was causing the problem. After making a Flight Safety Office contact to the base Naval Aerospace Engineering Office (NASO) with my findings, a NASO Representative was on site later that same day. The representative verified my findings and about five days later issued a Naval Aviation Safety Bulletin (NASB) for correcting the issue.

The NASB called for replacing the BDHIs with the longer canisters with BDHIs having shorter canisters. The BDHIs were the same basic part number; just a different ending dash number. The dash number indicated the longer canister – that was the only difference! The NASB further called for inspection of the associated BHDI connector wiring bundle, repair if necessary, and wrapping and tying of the wiring bundle. The NASB further referenced GenSpecs for the routine repair as needed, and I was given partial credit for this issue.

Another safety-of-flight issue arose and involved the installation of a wiring clamp around a wiring bundle near the pilot's throttle quadrant of an A-4M Aircraft. A fellow Aircraft Electrician and I were working on the issue at the same time. We brought the issue to the attention of the base NASO and they generated a GenSpec within one week for correcting the discrepancy. Because of my unique obligations at the time, this particular GenSpec was attributed to the fellow Aircraft Electrician. I did not mind, though, for the issue was safety of flight, and I did receive some amount of 'mentionable credit.'

The other three GenSpecs involved installing wiring clamps around lighting fixtures (in lieu of replacement); installing different sized data plates; and installing high-temp wiring clamps instead of regular wiring clamps. All of these GenSpecs involved the use of readily available government and industry materials in a different (but approved), application. The approved new applications saved the U.S. Government many thousands of dollars. And, for those two GenSpecs involved with Safety of Flight, they may have helped save some lives of pilots, ground crews, and possibly many others.

GenSPecs may be general in original nature. However, they can have *specific uses* for attaining far-reaching end goals on a wider scale than we ever thought possible. While traveling the course of a career and sojourning the pathways of one's life, the generalities within its reach should *never be considered a limited domain*. Looking outside the box of life's general specifications can change an existence...

...it can actually change the far-reaching world of destiny.

CHAPTER IV
BODY ARMOR

"Though I walk through the valley of the shadow of death...ya know God, we still could use some body armor around here right about .... now!"

Personal body armor (sometimes spelled armour) is defined as the whole of protective clothing, designed for absorbing and/or deflecting, slashing, bludgeoning, and penetrating attacks. Body armor was historically used for protecting soldiers. In modern times, body armor can also be employed for protecting police forces, security guards, bodyguards, and private citizens.

Body armor of antiquity ranged from primitive wood and stones to more advanced layers of extra clothing and toughened leather. As the ages passed, armor migrated to hammered metal plates and sheets of wire screens (chainmail). More advances in armor were made and were forged from carbon-steel plates, and in today's times are from composites of various super metals and fibers. Multiple factors have affected the development of body armor throughout human history. Significant factors in their developmental history included the economic and technological necessities for the logistics of armor production.

Interestingly, *all modern militaries usually* do not provide the best armor to their forces since doing such *would be very expensive*. There are times in the funding cycles of military weapon systems when the development of body armor will run parallel with the needed development of increasingly more effective weaponry on the battlefield. Thus, dilemmas are created for the manufacturers of body armor: it becomes effective armor versus effective weapons for the available funding.

The oldest known western armor dates from around 1400 BC and comes from the Mycenaean Age of Greece (around 1900 BC to 1100 BC). The armor was called Mail. The Mail was made from sheets of mail and interlocking iron rings which were either riveted or welded closed. In East Asian history, there was a similar style laminated armor called Lamellar, which was a coat of plates wired together (and was commonly used). Leather armor was also used which was made from rhinoceros hide.

Small plates of iron were eventually added to the Mail for protecting vulnerable areas. Thus, by the late 13th century, the knees, shoulders, and underarms became capped with armor. Various methods for improving protection were used as armor makers continued their many experimental designs. Into the 13th and the 15th centuries, iron armor could be carburized (case hardened) for giving its surface one of harden steel. Plate armor became cheaper to produce than Mail by the 15th century as it required much less labor. Mail continued to be used for protecting the joints which could not be adequately protected by plate armor.

With the advent of carburized iron, the small skull cap (helmet) evolved into the larger, true helmet; it was lengthened downward for protecting the back of the neck and the sides of the head. Several new forms of fully enclosed helmets were also introduced in the late 14th century which replaced the great helmets previously used in the Medieval Era with Renaissance Knights.

In the early 15th century, the small hand cannon (harquebus) first began to be used in the Hussite Wars, permitting infantry troops to defeat armored knights on the battlefield. About this same time in history, crossbows were being manufactured for being more powerful at piercing armor. Thus, what followed was about a one hundred and fifty year period in which better and more advanced steel armor was developed and used; specifically because of dangers posed by steel-pointed arrows and guns. In the early years of pistol and harquebus use, firearms were usually low in velocity. Full suits of armor and breast plates could actually stop bullets fired from a reasonable distance. It became common for the front armor breast plates to actually be used as shot tests. The impact points would be encircled and engraved for pointing them out as proof of impact. Armor often bore an insignia of the maker, especially if it was of good quality.

Surprisingly, instead of plate armor becoming obsolete, the use of firearms stimulated its development. While still being the targets of defending gunmen, armored horsemen were enabled to fight without being easily killed. Full suits of armor were also worn by generals and significant commanders up to the second decade into the 18th century. Though body armor was heavy, it permitted mounted commanders to survey battlefields while providing safety from gun fire.

Metal body armor still remained in limited usage long after its generally thought of extinction. Union and Confederate soldiers in the American Civil War obtained iron and steel vests from vendors and peddlers. Both the Union and the Confederate militaries had considered body armor, but rejected it for standard issue. Some of both Union and Confederate combatants obtained various styles of body armor with their own funds.

During the American Civil War, the two largest companies for manufacturing body armor were

G&D Cook Company and Atwater Armor Company. Both companies started the mass production of bulletproof vests when the Civil War first began. Both companies were located in New Haven, Connecticut. G&D Cook Company was originally a carriage maker, and was the largest maker of horse-drawn vehicles in the U.S. Not much information about the Atwater Armor Company is available, other than the fact that the company produced two hundred sets of body armors a day at its peak production rate during the Civil War.

The body armors manufactured by both companies were not manufactured to military specifications and were never officially issued to the troops on either side. There were no regulations or guidelines in place for controlling their designs and evaluations. Thus, both companies employed their own fabrications.

The G&D Cook Company manufactured two models, one for the infantry and one for the cavalry and artillery. The cavalry and artillery vest was the heavier model. It is not known if the Atwater Company manufactured more than one model, but it is known that their vests were heavier and more expensive than the ones from G&D Cook. Atwater's vest design used an added belt around the waist for a tighter fit.

The armor vests were widely advertised in the *Harper's Weekly* and *Frank Leslie's Illustrated Newspaper*. Some ads contained endorsements from military officers for selling them. Stores sold the vests for $5.00 for enlisted and $7.00 for officers (with higher quality ones for the officers). Effectiveness of those non-military obtained armor vests varied widely. Some of those armor units successfully deflected bullets and saved lives. Other armor units were poorly constructed and resulted in misfortune for the soldiers. In many instances, the obtained vests were abandoned by soldiers due to their weight on long marches. Also in many instances, there arose a stigma from fellow soldiers for the wearers of body armor being cowards.

At the start of WWI, thousands of the French Cuirassiers rode into battle engaging the German Cavalry who used steel helmets and steel armor. By then, shiny armor plating was covered in dark paint and a canvas wrap covered their elaborate Napoleonic-style helmets. Their light armor was intended for protection only against swords and daggers. The cavalry, unlike the foot soldiers who at least had trenches for protection, had to beware of high velocity rifles and machine guns.

Throughout the history of armed conflict, all combatants have desired to be protected as much as possible from swords, knives, spears, bullets, et cetera – aka *bulletproof.* That desire was, and still is, so strong that combatants will take extra measures for placing additional layers of protection between themselves and the danger zones of death. A very good case in point is the example of non-issued body armor during the American Civil War.

Yet there will never be any combatant who will be *100% bulletproof* from death's danger zones of war. But, we all still want that unobtainable guarantee.

And so, governments and agencies spend more and more funds for research and development in new and better methods of minimizing sustainable body injuries in the battlefield. From forged carbon-steel plates, that bulletproof desire carried combatants over the threshold with armor made of composites with various super metals. From that tactical advantage, the bulletproof desire has carried combatants over another threshold into the realm of threaded and woven silicon fibers – an even stronger and lighter armor than ever before. Those lightweight woven fibers have gone on to be manufactured in body armor of light-weight 'clothing' nature. Such technology has come a very long way from the ancient days of wood, stones, leather, and heavy metal plates.

It may be with the advent of ever increasing advances in laser technology that over the not-too-distant horizon, there will be 'shields of armor' like force fields such as those portrayed in the science fiction novels and movies. This concept can be stated because of the technology which already exists on a limited basis for the eyes of aircraft pilots – it is called '*Laser Armor.*' Lasers have been employed (for a number of years) by the military for shooting down missiles and other targets. And thus, from those concepts of 'laser armor protection' one day laser body armor ***WILL*** be a reality.

MilSpecs were never in the early beginnings of body armor (or vehicle armor). As the years passed and standardization of processes was required for achieving success, MilSpecs were applied to the logistics of acquiring armor equipment and maintaining those assets within the military supply systems.

Before leaving this entry on Body Armor, let us take a look at another application of armor – a spiritual one. Every human being's enemy (especially the Christian) is the devil (satan), and he will oppose us all. Most of his victims are never even aware there is a spiritual warfare going on in their lives. Therefore, they fall a very easy prey to the forces of evil. Christians should become very perceptive to the fact that we are involved within a great spiritual struggle, even though many never consider themselves in a spiritual war with satan. They do not understand the nature of his schemes and the weapons he employs – or of the weapons which God has provided for our defense.

Nevertheless, there is an ongoing spiritual warfare -a mystical conflict- within the paranormal realm over every soul on this planet.

In Ephesians 6:10-18 is possibly the clearest definition of the spiritual warfare which can be found in the Holy Bible, and it was penned by the Apostle Paul. Its passages not only declare there is actually a spiritual war, but its words warn us all that, apart from utilizing the weapons which God has provided for us, we are hopelessly underpowered. These passages of Scripture inform us as to what those Divinely God-given weapons are. Those Scriptures imply the nature of the spiritual struggle in which every Christian is enthralled. They also delineate the weapons which God has provided for repelling the attacks of satan. Thus, by reading and studying those passages of Scripture, we can learn much about satan's oppositional strategy. We can also discover that God has our back in times of conflict – if we will yield to God, and be prepared with the provisional armor from Heaven. Ephesians 6:10-18 reads as follows:

"Finally, my brethren, be strong in the Lord and in the power of His might. Put on the whole armor of God, that you may be able to stand against the wiles of the devil. For we do not wrestle against flesh and blood, but against principalities, against powers, against the rulers of the darkness of this age, against spiritual hosts of wickedness in the heavenly places. Therefore take up the whole armor of God, that you may be able to withstand in the evil day, and having done all, to stand. Stand therefore, having girded your waist with truth, having put on the breastplate of righteousness, and having shod your feet with the preparation of the gospel of peace; above all, taking the shield of faith with which you will be able to quench all the fiery darts of the wicked one. And take the helmet of salvation, and the sword of the Spirit, which is the word of God; praying always with all prayer and supplication in the Spirit, being watchful to this end with all perseverance and supplication for all the saints -"

It pays to have the full 'spiritual armor' of God as *clothing* on us at all times. We need to be prepared for whatever the enemy of our soul throws against us. Within ourselves, we can never win over a spiritual giant from the spirit realm. However, with God and His spiritual armor *provisioned especially for us*, we can win those impossible spiritual battles

– and eventually win the war of eternal security for our soul!

Left – Body Armor of the Middle Ages
Right – WWI German Body Armor

Military Body Armor of the 21st Century

CHAPTER V
SMALL ARMS

"Ya know Sarge, right about now I could really go for one'a them Tommy Guns. Ya seen any around here lately?"

From spears, to slings, to slingshots, to bows and arrows, to catapults, combatants have endlessly searched for new, better, and rapid ways of launching devastating projectiles at their enemies. And so, around the early beginnings of the 6th century, scientific inventors in ancient China began formulating varying grades of gunpowder along with experimenting with different types of fireworks. Their explosive results launched designs of single and multiple-launch artillery rockets, single-shot smooth-bore fire lances, and multi-barreled guns. Their quest for volatile engagement also eventually produced the world's first cannon, which (oddly) was made from cast bronze!

Centuries later, in the late Dark Ages of Europe, the term 'firearm' was used in Old English for denoting the arm in which the match was held that was used to light the touch hole on the volatile hand cannons. The term was a variation from the then contemporary terms used for 'bow arms' and 'drawing arms' used in archery. Because of the excessive blast effects from firing the gun barrel at the time, the operating gunner was more safely located somewhere behind the gun, while steadying it with the other hand. Thus eventually, the term 'firearm' became synonymous with the term 'hand gun' which is still used today for small portable guns.

Back in China, so *prolific* became their firework creations, and with them becoming portable into the form of firearms and handguns, that Chinese citizens began owning them by the common masses. This practice became viewed as a seriously grave threat to the Chinese government.

As a result, the world's first gun control law appeared from the direction of the Shogun of China, and was as written follows:

'The people of the various provinces are strictly forbidden to have in their possession any swords, bows, spears, firearms, or other type of arms. The possession of these elements makes difficult the collection of taxes and dues and tends to permit uprising, therefore, the heads of the provinces, official agents, and deputies are ordered to collect all weapons mentioned above and turn them over to the government.'

Toyotomi Hideyshi, Shogun
29 August 1558

The earliest firearm depiction is a sculpture from a cave in Sichuan, China, and dates back to the 12th century. The sculpture is of a figure carrying a vase-shaped bombard (cannon) with flames and a ball being ejected from its opening. The oldest known surviving gun from antiquity is made of bronze, has been dated to the year 1288, and was discovered at an excavation site in Heilongjiang, China. The Europeans, Arabs, and Koreans obtained firearms around the 14th century. The Turks, Iranians, and Indians obtained firearms about a century later (directly or indirectly from the Europeans). The Japanese did not acquire firearms until the 16th century, and then (surprisingly) it was from the Portuguese and not from the Chinese.

Development of firearms greatly accelerated during the 1800s and 1900s. It was during that era when firearm breech-loading technology became somewhat of a universal standard for the reloading of most firearms. From loading individual rounds into weapons, magazines holding multiple munitions were developed and rapid reloading simply became a fundamental *requirement* in conflict. Rapid reloading then paved the way for concepts of semi-automatic automatic firing mechanisms. This development meant soldiers could fire more rounds per minute than older weapons could fire throughout the course of battles.

Metal alloys to composite materials and polymers in firearm construction made overall weaponry becoming progressively stronger and lighter. Ammunition evolved over the centuries from simple metallic ball-shaped projectiles rattling down smooth bore barrels, to bullets and cartridges manufactured to high tolerances ejected at high speed from grooved (rifled) barrels. Especially since the early 20th century, particular attention has been devoted to accuracy and sighting for making firearms far more accurate than has ever been before. One salient factor for proliferating firearms has been due to the advent of mass production – the same as with any mass produced item.

There have been many firearms which have greatly impacted the military. However, there were perhaps only four firearms which changed the face of the military into the 20th century for decades. Those firearms are the '45 Auto' (M1911 Pistol), the 'Tommy Gun' (Thompson Submachine Gun), the 'AK-47' (AK-47 Assault Rifle), and the 'M-16' (M-16 Assault Rifle).

The M-1911 Pistol is a semi-automatic American Pistol. It is magazine-fed, recoil-operated, and chambered for the .45 ACP cartridge. John M. Browning invented and designed the firearm and it became the standard-issue side arm for the U.S. military from 1911 to 1985. At the time of writing this book, the M1911 Pistol is still carried by some U.S. military forces.

Battlefield weaponry problems had been encountered by American units fighting the Moro guerrillas during the Philippine-American War. The then-standard Colt M1892 revolver (using .38 Long Colt caliber ammunition) was found unsuitable for jungle warfare, especially for stopping power of the enemy. The Moros had very high battle morale and, being frequent drug users, did not feel much pain in combat. The U.S. Army briefly reverted to using the M1873 single-action revolver in .45 Colt caliber ammunition. It had had been standard during the late 19th century; and the heavier .45 caliber bullet was found more effective against the charging hyped-up Moros tribesmen. The .38 Long Colt problems led the U.S. Army to shipping new single-action .45 Colt revolvers to the Philippines in 1902. This large supply action prompted the then Chief of Army Ordnance to authorize testing for a new Army Service Pistol.

In 1904, following the Thompson-LaGarde Pistol Round Effectiveness Tests, Colonel John T. Thompson issued a statement that the proposed new service pistol should not be of less than .45 caliber and would preferably be semi-automatic in operation. This statement then led to the 1906 pistol trials with the contest finalists of firearms manufacturing companies: Colt Firearms, Bergmann, Deutsche Waffen und Munitionsfabriken

(DWM), Savage Arms Company, Knoble, Webley, and White-Merril.

Of the final designs submitted, only three remained for careful consideration: the Savage, Colt, and DWM designs (all chambered with .45 ACP cartridges). These three pistols still had design issues needing correction; however, only Colt and Savage resubmitted their designs. A series of field tests from 1907 to 1911 were conducted for deciding between the Savage and Colt designs. Both designs were improved between each testing over their initial entries, leading up to the final test before adoption.

For Colt's test at the end of 1910, the test was attended by its designer, John Browning. Over the course of two days, 6,000 rounds were fired from a single pistol. When the gun began growing hot, it was simply immersed in water for cooling it down. The Colt gun passed with no reported malfunctions, while the Savage gun design had thirty-seven reported discrepancies.

Following its successful trials, the Colt pistol was formally adopted by the U.S. Army on 29 March 1911, thus gaining its formal U.S. Army designation as: Automatic Pistol, Caliber .45, M1911 (in 1924 after some modifications, the M1911A1). It was adopted by the Navy and Marine Corps in 1913. Originally manufactured only by Colt, demand for the firearm in WWI saw rapid expansion of manufacturing into the government-owned Springfield Armory.

To date, the M1911 Automatic Pistol is used by twenty-eight nations and is made by a wide variety of arms manufacturers. It was widely used in WWI,

WWII, the Korean War, and the Vietnam War. The U.S. procured around 2.7 million M1911 and M1911A1 Pistols in military contracts during its prime service use life.

The beginnings of its long and glorious military career with this distinguished pistol began with *only one specification* – the one issued by Colonel John T. Thompson; a statement that the new service pistol should not be of less than .45 caliber and that it should preferably be semi-automatic in operation. There were a few more specifications added along the developmental pathway of its logistical process – but very few for a great success which won many battles and two world wars.

Never underestimate what the power of a great vision can do.

The Tommy Gun is an American .45 caliber automatic rifle invented by John T. Thompson in 1919. The weapon became very infamous during the American Prohibition Era of the 1920s. It was a common sight in the media of that time, being well used by both law enforcement officers and criminals alike. The Thompson Gun was also informally known as: 'Trench Broom,' 'Trench Sweeper,' 'Chicago Typewriter,' "Chicago Piano,' 'Chicago Style,' and 'The Chopper.' The 'Tommy Gun' began being favored by soldiers, criminals, police, and civilians alike the world over for its compactness, reliability, high rate of automatic fire, and its use of the large .45 ACP cartridge.

In developing the Tommy Gun, General John T. Thompson originally envisioned a semi-automatic rifle for replacing the bolt action service rifles in use before 1916. While searching for ways of permitting such a weapon to operate safely without the complexity of recoil or gas operated mechanisms, Thompson came across a patent issued to John Bell Blish in 1915. Mr. Blish's patent employed a different principle and was based on adhesion of inclined metal surfaces under pressure. Thompson found financial backing and started the Auto-Ordnance Company in 1916 for the purpose of developing his new rifle.

The principal designers for the new rifle were Theodore Eickhoff, Oscar Payne, and George Goll. By late 1917, limits of the Blish Principle were discovered and it was found that the only cartridge currently in U.S. service suitable for use with the Blish Principle design was the .45 ACP round. Thompson then envisioned a 'one-man, hand-held machine gun' in .45 ACP caliber as a 'trench broom' for use in the on-going trench warfare of WWI. Payne designed the gun along with both the stick and drum type ammunition magazines. The project was then titled 'Annihilator I' and by 1918, most of the design issues had been resolved. However, WWI ended before prototypes could be shipped to Europe.

In an Auto-Ordnance Company Board meeting in 1919 to discuss the marketing of the Annihilator Machine Gun with the war over, the weapon was officially renamed as the 'Thompson Submachine Gun.' Other weapons had been developed just prior with similar objectives in mind, yet the Thompson Submachine Gun was the first weapon to be labeled and marketed as a 'Submachine Gun.'

General Thompson intended the weapon as an automatic 'trench-broom' to sweep enemy troops from the trenches, filling a role for which the Browning Automatic Rifle had been proven not well suited for.

To date, the Thompson Submachine Gun has been used by the military and police of at least twenty-five nations. With the Tommy Gun still in use today, undetermined millions of these machine guns have been made by various arms manufacturers. In the span of about three years, the Tommy Gun came into existence with some funding, one existing patent, four designers, a few specifications, and a tremendous vision.

Never underestimate what the power of a great vision can do.

The AK-47 Assault Rifle is a selective-fire, gas-operated 7.62×39mm caliber rifle designed for combative assaults. The assault weapon was first developed in the Union of Soviet Socialist Republics (USSR) by Mikhail Kalashnikov. It is originally officially known as Avtomat Kalashnikova (Автомат Калашникова in Russian). It is also known as a Kalashnikov, or an 'AK' which in Russian slang is Kalash.

During WWII, the Germans began pioneering the assault rifle concept, which was based on research revealing most firefights occur at close range (usually within about 325 yards). Power and range requirements of most rifle cartridges were excessive for the majority of small arms firefights. These facts resulted in the armies seeking a combination of cartridge-rifle-submachine-gun features for their troops in the field. Additionally, the small arms weapon should have large-capacity magazines, selective-fire capability, and an intermediate-power cartridge effective to 325 yards. In order for reducing immediate manufacturing costs, the 7.92x57mm Mauser cartridge case was shortened, the result of which was a lighter 7.92x33mm Kurz cartridge.

From all this research and development, its results birthed the Sturmgewehr 44 (StG44) Assault Rifle. Toward the end of WWII, the Germans fielded the StG44 against the Russians, which most certainly its impacting field experience notably influenced the Russian military, especially during its post-war years.

In September 1941, Mikhail Kalashnikov was a Senior Sergeant and Tank Commander serving on the T-34 Tanks with the 24th Tank Regiment, 12th Tank Division of the 8th Mechanized Corps, and stationed in Stryi, Ukraine (in the foothills of the Carpathian Mountains). Troops of the German XXXXVII Corps crossed the Desna River in the area southwest of Moscow, Russia. During the invasion attack, the German troops broke through front lines at the City of Bryansk, thus beginning the Battle of Bryansk on 30 September 1941. During the battle, the City of Bryansk was subjected to very heavy artillery and air bombardment which destroyed large parts of it. The Soviet Red Army, as part of the Orel-Bryansk Defensive Operation (30 September to 23 October 1941) lost over 80,000 as KIA in the engagement. Another 50,000 prisoners of war (POWs) were taken by the German XXXXVII Corps before proceeding on to their next objective – Orel, another city southwest of Moscow.

Sergeant Kalashnikov was wounded in the shoulder during the Battle of Bryansk. Though wounded and traveling on foot, he was able to make it to a hospital where he received medical attention. This was not Kalashnikov's first battle; he also participated in the Battle of Brody where his division encountered the German 16th Panzer Division in battle around Leshuv. While he was hospitalized in Bryansk, Kalashnikov ended up becoming very sick and received a six-month release from his duties for recuperation because of his illness. While in the hospital, he overheard some fellow soldiers complaining about the Soviet rifles. Knowing he also had difficulties with the standard infantry weapons, he decided to experiment with constructing a new rifle for the Soviet military. The main catalyst, though, for his experiments of finding a new rifle design was his flashbacks. It was while being hospitalized and recovering from his injuries that Sergeant Kalashnikov started experiencing flashbacks of enemy raids. With those flashbacks he 'became obsessed with creating a submachine gun that would drive the Germans from his homeland.'

It was there in a hospital recovering from being shot, that Sergeant Mikhail Kalashnikov actually began a successful new career as a Small Arms Weapons Designer for the Soviet Union. Sergeant Kalashnikov did, in fact, design a new weapon. Although his first submachine gun design was not accepted into service, his talent as a weapons designer did not go unnoticed. From 1942 an onward, Sergeant Kalashnikov was assigned to the Central Scientific-Developmental Firing Range for Rifle Firearms of the Chief Artillery Directorate of the Рабоче-Крестьянская Красная Армия – the PKKA (or the Workers' and Peasants' Red Army). Simply stated, it was the Red Army. In English the PKKA was transliterated as Raboče-Krĕst'janskaja Krasnaja Armija – or the RKKA.

A few years later in 1944, after tinkering with submachine gun designs, he entered a competition for a new weapon that would chamber an M1943 7.62x41mm cartridge. One especially particular requirement of the competition was the reliability of the firearm in the mud, damp, and frozen conditions of the Soviet front line. Kalashnikov had designed a carbine, strongly influenced by the American M1 Garand, but lost out to the Simonov design (a scaled down PTRS-41 that would later become the SKS Assault Carbine).

At the same time, the Soviet Army was interested in developing a truly portable assault rifle which was capable of firing a shortened M1943 cartridge (7.62x39mm). The first such weapon was presented by Sudayev in 1944, but trials found it being too heavy. Another competition was held two years later for new designs, and there Kalashnikov and his design team submitted an entry. It was a gas-operated rifle which had a breech-block mechanism similar to his 1944 carbine, and a curved 30-round magazine.

Kalashnikov's rifles were codenamed AK-1 and AK-2 and proved most reliable as the weapon was accepted into the second round of competition. Weapon designs from Demetev and Bulkin were also accepted. In late 1946, as the rifles were being tested, one of Kalashnikov's assistants suggested a redesign for the AK-1 for improving its reliability. The improved rifle was produced for a second round of firing tests and field trials. There, Kalashnikov's AK-1 Model 1947 Assault Rifle proved to be simple and reliable under a wide range of conditions with simple handling uniqueness. In 1949 the rifle was adopted by the Soviet Army as the '7.62mm Kalashnikov Assault Rifle (AK).'

Since 1949, the AK-47 Assault Rifle has been used by over seventy-four nations, and has been produced in hundreds of variants from its original design. Since 1949, approximately seventy-five million AK-47 rifles have been produced. Additionally since that year, approximately one hundred million AK-47 variant rifles have been produced. The AK-47 Assault Rifle is still commonly used throughout the world today.

In the year of 1949 when the AK-47 became the new small arms weapon of the Soviet Army, Mikhail Kalashnikov resided and worked in Izhevsk, Udmurtia. As the time of the writing this book, Kalashnikov still resides in that city. He is age ninety-two and holds an Advanced Degree in Doctor of Technical Sciences. He was twice named Hero of Socialist Labor. In 1998 he was awarded an Order of Saint Andrew the Protoclete. On his 90th birthday on 10 November 2009, Kalashnikov was named a 'Hero of the Russian Federation' and presented with a medal by President Dmitry Medvedev who praised him for creating 'the brand every Russian is proud of.'

Despite estimates of some one hundred million AK-47 assault rifles still in circulation, General Mikhail Kalashnikov claims he has not profited from its invention, and that he only receives a state pension. Research will show that sources claim he does, however, own thirty percent of the Marken Marketing International (MMI) which is based in Solingen, Germany. The company revamps trademarks and produces merchandise carrying the Kalashnikov name, such as vodka, umbrellas, and knives.

By all appearances, Kalashnikov seems to be a humble man. He is quoted as saying, 'It is painful for me to see when criminal elements of all kinds fire from my weapon...I created this weapon primarily to safeguard our motherland.' After the War on Terrorism began in 2001, Kalashnikov was quoted as stating in 2002, 'When I see Bin Laden with his AK-47, I got nervous. But what can I do, terrorists aren't fools: they too chose the most reliable guns. I'm proud of my invention, but I'm sad that it is used by terrorists.' Later in life, Kalashnikov claimed that the priority of simplicity and dependability in his designs was influenced by principles he had gained from reading Russian literature and the Holy Bible.

Never underestimate what the power of a great vision can do.

The M-16 Assault Rifle is a lightweight, 5.56mm caliber, air-cooled, gas-operated, magazine-fed selective fire assault rifle. The weapon is made of steel, 7075 aluminum alloy, composite plastics, and polymer materials.

In 1948, the U.S. Army organized the Civilian Operations Research Office, which paralleled similar research operations organizations in the United Kingdom (UK). One of its first efforts was Project ALCLAD; a study of body armor. The project's conclusion was that more information was needed about battlefield injuries in order for arriving at reasonable conclusions. Therefore, over three million battlefield reports from WWI and WWII were analyzed during the next few years before a series of reports on the project's findings were released.

Some of the report conclusions suggested that infantry should be equipped with fully-automatic rifles of some sort which would increase the actual firepower of regular soldiers. Additionally, another very clear fact was automatic rifles increased ammunition use and, thus, the infantry would be required to carry more ammunition into firefights. Therefore, the infantry soldier should have a lighter rifle with light-weight ammunition. Findings also were that few existing rifles met little of those criteria.

Although it appeared the new 7.62mm T-44 Rifle (precursor to the M-14 Rifle) would increase the rate of fire, its heavy 7.62mm NATO cartridge made carrying significant amounts of ammunition cumbersome. In addition, the length and weight of the T-44 Rifle made it unsuitable for short range combat situations often found in jungle combat, mechanized warfare, urban conflict, or wherever a smaller and lighter weapon was needed for being brought to bear faster in use.

A new study was launched (Project SALVO) which was set up for researching a weapon design suited to real-world combat. Operating from 1953 to 1957 and in two phases, Project SALVO eventually suggested that an automatic weapon firing four rounds into a 20-inch area would double the hit capability of existing semi-automatic weapons. In the second phase (SALVO II), several experimental weapon concepts were tested in the .22, .25, .308, and .30-06 caliber cartridge range.

In 1954, Mr. Eugene Stoner of the newly-formed ArmaLite Company helped develop the 7.62mm AR-10 Rifle. ArmaLite was founded specifically to bring the latest in designs and alloys to firearms design, and Mr. Stoner felt he could easily exceed the other designs. In 1957, not long after a composite barrel burst with one prototype, the AR-10 Rifle was rejected. The AR-10 was later produced in Europe and saw limited, but very successful, military service with several foreign nations.

In 1957, Mr. Stoner designed and built the AR-15 Assault Rifle. It was a lightweight, 5.56mm, selective-fire, air-cooled, gas-operated, magazine-fed, rotating-lock bolt, automatic rifle. The rifle was manufactured with extensive use of aluminum alloys and synthetic materials. Even though it was a very impressive weapon, ArmaLite sold the AR-15 and AR-10 designs to Colt Firearms in 1959. The sale was because of Armalite's financial problems.

After Colt provided a tour of the AR-15 weapon in the Far East, the first sale of AR-15s was made to

Malaysia in September 1959 with Colt's manufacture of their first three hundred AR-15s in December 1959. Colt then marketed the AR-15 rifle to various military services around the world, which also included the U.S. military. However, Colt continued to use the AR-15 trademark for its semi-automatic variants (AR-15 and AR-15A2, which were marketed to civilian and law-enforcement customers). The original AR-15 was a very lightweight weapon, weighing less than six pounds with an empty magazine. Later heavier barrel versions of the civilian AR-15 would weigh upwards of 8.5 pounds.

In July 1961, the U.S. Air Force ordered 80,000 AR-15s, and some of those were designated for service in South Vietnam. The next year, another one thousand AR-15s were sent to South Vietnam and distributed among Allied Forces. Special Operations units and advisers working with the South Vietnamese troops filed battlefield reports abundantly praising the AR-15. Additionally, they widely acclaimed the stopping effectiveness of its .223 (5.56mm) cartridge and pressed for adoption of the AR-15 into military service. What no one knew, though, except the men directly using the AR-15s in the field, were the destructive impacts made by the new weapon. Photographs showing results of enemy casualties made by the .223 round remained classified on into the 1980s. What made the .223 bullets so devastating were its vast speed and the fact that it tumbled on its trajectory. When impacting a target, the point of impact was generally decimated.

By 1963, the AR-15 Assault Rifle was re-designated as the M-16 Assault Rifle. In November

1963, the U.S. Secretary of Defense approved the U.S. Army's order of 85,000 XM-16E1s for jungle warfare operations. The U.S. Air Force then ordered another 19,000 M-16s. The U.S. Army immediately began issuing the XM-16E1 to infantry units; however, the rifle was initially delivered without adequate cleaning supplies or instructions. This oversight generated grave problems on the battlefield, resulting in hundreds of soldiers losing their lives. Once the cleaning kits were issued and proper field cleaning was maintained, the M16 Assault Rifle was a field-ready weapon once again.

The M-16 Assault Rifle replaced the M-1 Carbine and M-14 Rifle as standard infantry rifles for the U.S. military. The M-14 Rifle continues limited service; mostly in sniper, designated marksman, and ceremonial roles. The M-16 Assault Rifle is the most commonly manufactured 5.56x45mm rifle in the world. The M-16 Assault Rifle is used by fifteen NATO countries and more than eighty countries worldwide. In addition to America, numerous weapons manufacturing companies around have produced more than eight million M-16 Assault Rifles of all variants. Approximately ninety percent of them are still in operation.

That great mix of visionaries brought four very successful new small arms weapons to the world during times of crisis. They helped win multitudes of battles, two world wars, and their fame is known the world over.

Never underestimate what the power of a great vision can do, especially if God is recognized in the design work – if in doubt, just ask General Mikhail Kalashnikov.

M-1911 U.S. Army Colt Automatic Pistol

M-1928 U.S. Army Thompson Machine Gum

Soviet AK-47 Assault Rifle

M-16 U.S. Army Assault Rifle

CHAPTER VI
STEEL POTS

"If that new guy don't quit pointin' that gun our way, we're gonna have ta start wearin steel pots around here!"

Helmets are one of the oldest forms of personal protective equipment against the harshness of combat. They were known to have been worn by the Assyrians around 900 BC and have continued in their widespread use by many militaries ever since then.

Materials and construction for combat helmets became more advanced as weapons became increasingly more powerful. Originally, helmet construction was from leather and brass, and then bronze and iron during the Bronze and Iron Ages. Around 950 AD, many societies began producing helmets from forged steel. Their main purpose, however, was a military one: *protecting the head from cutting blows with swords, flying arrows, and low-velocity projectiles.* Many medieval helmet designs rested on the wearer's shoulders, which prevented turning their heads and thus, greatly restricted mobility.

Because of the need for speed and mobility, the use of helmets in combat declined after 1670, and rifled firearms ended their wide use by foot soldiers after 1700. Instead of helmets, many armies used non-armored headgear, which rendered little protection from bullets and blades. The Napoleonic Era saw ornate cavalry helmets reintroduced for cuirassiers and dragoons in some armies which continued to be used by French Forces during WWI as late as 1915. WWI, with its trench warfare and wide use of artillery, brought a return to the mass use of metal helmets. This return, though, came with a shape offering not only more protection, but mobility, low profile, and compatibility with gas masks.

Up to the beginning of WWI, the Germans had an advantage with combat headgear. Many of their foot soldiers wore 'Pickelhauben' for combat headgear. The term Pickelhauben is derived from the old German word 'Pickel' meaning 'point' (or 'pickaxe'), and 'Haube' meaning 'bonnet.'

Pickelhauben were used in the 1800s by Prussians; and one design had a low back for protecting the neck of the wearer, which was commonly called a lobster tail. Pickelhauben originally came in hardened leather construction but, in the 1890s, were manufactured in steel. Pickelhauben were made for attaching a spike on the center of its top. This was for affixing a stream of horse hair and was for show. The spike could actually become another deadly weapon in close combat.

Even though the Germans had a tremendous advantage for headgear at the onset of WWI, most of their combatants were not issued with any form of special headgear protection in the beginning of war. Cloth and leather hats were mostly issued and those were basically designed for protection against saber and blade cuts. When trench warfare began, casualties on all sides escalated from severe head wounds; usually from shrapnel rather than from gunfire.

The Imperial German Army began researching designs of the Stahlhelm (steel helmet) which was accomplished by Dr. Friedrich Schwerd of the Technical Institute of Hanover. In early 1915, Dr. Schwerd had studied head wounds suffered during trench warfare and submitted a recommendation for steel helmets. Soon after being transferred to Berlin, he began the task of designing and producing a suitable helmet broadly based on the 15th century sallet (helmets with face visors), which provided good protection for the head and neck.

After a lengthy and in-depth developmental process (including testing selections of both German and Allied headgear), the first Stahlhelms were tested in November of 1915 at the Kummersdorf Proving Ground and then field tested by the German First Assault Battalion. Thirty thousand examples were ordered, but it was not approved for general issue until after the New Year of 1916. The new German helmet became referred to as the 'German Stahlhelm Model 1916' or simply 'M16 Stahlhelm.' In February of 1916, the M16 Stahlhelm was distributed to the German troops at Verdun; its issuance *dramatically lowered* their incidence of serious battlefield head injuries.

Instead of the steel composition used for the newly developed British Brodie Helmet, the Germans employed a harder alloy steel for theirs – a silicon-nickel-steel alloy. The helmet's basic shape was that of a 'coal shuttle' (for scooping coal). These factors resulted in the new helmet having to be formed within heated dies at higher unit costs than the British helmet.

The M1916 Stahlhelm design had horn-like ventilator lugs installed on each side of the helmet. These served for supporting the addition of a steel brow plate (Stirnpanzer). The brow plate provided additional ballistic protection from bullets and fragmentation. Because the brow plate was too heavy for general use, it was typically limited to use by snipers and trench raiding parties. The M1916 Stahlhelm, though, was not without its shortcomings. The ventilator horns would allow cold air in during the winter, requiring the wearer to block the vents with whatever was available. The helmet's large flared skirt tended to generate difficulties with hearing, often distorting the surrounding sounds and creating echoes.

Originally the German M16 Stahlhelms were painted Feldgrau (field grey) in color, and they were often camouflaged by troops in the field using mud, foliage, cloth covers, and paint. Official issued cloth covers in white and grey appeared in late 1916 and early 1917, with official camouflage paint being issued in July 1918. After the effectiveness of the M1916 Stahlhelm design was validated during the 1916 battle campaigns, small improvements were suggested and implemented. The M1917 design had improvements to its headliner, but otherwise was identical in all other aspects. Another improvement came to the M1918 Stahlhelm; its shell did not have the chinstrap rivets on the lower side of the helmet skirt which were found on earlier models.

The German M1916 Stahlhelm design proved to provide excellent protection on the battlefield. It went on to become the standard issued helmet for the German Army until 1935 (when it was redesigned). From 1935 through WWII, Germany designed and issued Stahlhelm Models M35, M40, and M42 for combat use. The M16, M17, and M18 Stahlhelms still remained in limited use. Those six models of German Stahlhelms were used by over twenty-six nations for their military and police forces, with millions of them being produced by various manufacturers.

As previously stated, when trench warfare of WWI began, casualties on all sides escalated from severe head wounds, and it was usually from shrapnel rather than from gunfire. The French, though, were actually the first to observe a serious need for more protection and began experimenting with steel headgear. In 1915 the French invented and introduced the M15 Adrian Helmet for their soldiers, and it *became the* ***first steel combat helmet*** *for modern warfare of its time*.

In England, the British War Office had also observed the serious need for steel helmets. The War Office Invention Department evaluated the M15 French Adrian helmet, but determined its design was not strong enough and was too complex for being rapidly manufactured at that time. A helmet design by John Brodie of London, patented in 1915, offered significant advantages over the French Adrian design, because (1) it was constructed from a single unit which could be pressed from a single sheet of steel (providing it added strength), and (2) the grade of sheet steel was stronger.

The English Brodie helmet's 'soup bowl' shape with its wide brim was designed for protecting the wearer's head and shoulders from falling fragmentation. Although this design, coupled with being pressed from one piece of sheet steel, made it more resistant to projectiles, the design offered less protection for the lower part of the head and neck than other designs. The Brodie Helmet Type A was in production only a few weeks before being modified to Type B in October 1915, when the British and Commonwealth troops had begun being issued the helmets. Soon afterwards the English troops nicknamed them the 'Tommy Helmets.' In 1916 the British War Office required the helmets be manufactured from non-magnetic manganese steel and thus, the British Mark I Brodie Helmet came into existence. Between September 1915 and February

1919, more than seven million British Brodie Helmets were made, issued, and also exported.

In the pre-war days of WWI, America was trying to stay out of the war – but knew it was actually just a matter of time before being drawn into its swelling conflict. At the time, the U.S. military was experimenting with several designs of steel helmets, but did not have one ready for mass production and issue. Therefore, the U.S. military adopted the British Mark I Brodie Helmet; and the first units to receive issue were the American Expeditionary Forces (AEF) operating in France in July of 1917.

In the fall of 1917, the U.S. began production with its own design of combat helmets; they were nearly identical to the British Mark I Brodie Helmet. The composition was slightly different with manganese steel, its coating was with saw dust, and its color was painted olive drab. The American GIs of that time period were referred to as 'Doughboys.' Thus, the American GIs called their new combat helmets, 'Doughboy Helmets.' With these helmets appearing all over Europe, an interesting media article appeared in the *Stars & Stripes Magazine*, on Friday, 3 May 1918, which partially read as follows:

> IT'S THE STEEL STETSON
>
> They're calling them the "Steel Stetsons" now....but it's a lot safer than the average Stetson for wear in the northern most regions of France where the sky is full of a number of things all day and all night. So "Steel Stetsons" they are until somebody else can think up something better.

A helmet was designed by Major Harold Sydenham in December of 1940 and it became known as the 'TS3 Experimental Helmet.' This helmet, along with the M1917 Doughboy Helmet and a variety of other helmets, underwent testing at Fort Benning, Georgia. The TS3 surpassed all the other helmets in field testing, and was designated as the U.S. Military Number 1 Helmet by the U.S. Army Infantry Board in February 1941. The M1 Helmet became standardized for military issue on 30 April 1941 and was approved for manufacture on 9 June 1941.

The M1 Helmet was of a two-piece design with an outer manganese steel shell and a separate fiber inner headliner containing the suspension system. Originally, the steel shell was made by the McCord Radiator Company of Detroit, Michigan, and the fiber headliner assembly was manufactured by the Hawley Products Company. Because the M1 Helmet came in two parts, quickly separable, it became immediately popular with all the military. This creative design led to some novel uses when the steel shell was separated from its liner. The shell could be used as an entrenching tool, washbasin, bucket, seat, stool, and, in desperate times, even a latrine. The shell was also used as a cooking pot but this practice was discouraged, as it could cause the metal to become weakened over time.

Because of both the wide utility and the strength of protection offered in combat by the M1 Steel Helmet, the American GIs quickly nicknamed their favorite helmet the 'Steel Pot.'

Manufacture of the M1 'Steel Pot' Helmet began in June 1941 and ended forty-four years later in 1985. During those years, the M1 Helmet was the standard issued combat helmet of the U.S. military. From June 1941 through September 1945, over *twenty-two million M1 Helmets were manufactured in the U.S.* During 1966 to 1967, a second production run produced approximately one million more of these helmets for the Vietnam War. The M1 Helmet has been very popular for use in militaries and police use around the world in over one hundred nations since WWII. Many weapons companies in other countries have also manufactured this helmet. Thus, the actual total millions of numbers of these helmets having been produced are difficult to accurately determine.

For over forty-four years since its creation in 1941, the M1 'Steel Pot' Helmet became an American standard in combat helmets around the world – until composites come along.

The PASGT Helmet (Personal Armor System for Ground Troops) was developed in 1975, but did not see its first combat use until October 1983 during Operation Urgent Fury in Grenada. In 1985, the U.S military totally replaced the M1 Helmet with the PASGT Helmet. The PASGT Helmet is most commonly known as the 'Kevlar Helmet' and is also referred to as the 'K-Pot' (picking up on the term 'Steel Pot' from the M1 Helmet). The PASGT is also, but less commonly, referred to as the 'Fritz Helmet' for its resemblance to WWII German Helmets. The shell is made from multiple layers of Kevlar. Kevlar is a ballistic Aramid fabric, treated with a phenolic resin, and offers very high protection against shrapnel and ballistic threats.

At the time of writing this book, the PASGT Helmet is being phased out and replaced by the MICH Helmet (Modular Integrated Communications Helmet). Its composition is a new and more advanced type of Kevlar which provides increased protection against bullets and fragmentation. Its headliner has a pad retention system similar to the cushions and straps found on bicycle helmets. This replaces the nylon cord suspension system, sweatband, and chinstrap employed in the PASGT helmets. The MICH Helmet is scheduled to be standard issue for all the U.S. military service.

More militaries into early 21st century will begin using high-quality helmets made of ballistic fiber materials such as Kevlar and Aramid, which have excellent bullet and fragmentation stopping power. These helmets are commonly called 'Ballistic Helmets.' Some ballistic helmets also have good non-ballistic protective qualities against explosion concussive shockwaves, physical attacks, vehicle accidents, falls, et cetera.

From the Pickelhauben Helmet of the 1800s to the MICH Helmet of the early 21st century, very few MilSpecs were utilized. Most that were used were with the PASGT and the MICH. The others employed patents, contracts, and some industry standards. It was the use of those 'specifications' along with MilSpecs in the latter years around WWII and after, and also a drive for attaining goals, that brought needed effective protective headgear into the world's arena of armed conflict.

MilSpecs may not only narrow the search for testing and validation, but they can also provide an enlightened pathway toward broader bases of technology applications.

In spite of higher and greater technological advances, there will always be favorites in anything for obvious reasons. From the Pickelhauben Helmet of the 1800s, to the MICH Helmet of the early 21st century, came new technology and flashy designs over the years. With it the M1 Helmet remained a favorite, and it will remain a favorite for a very long time to come. It has endured battle after battle, war after war, year in and year out, for over seventy years – and it is still going strong in combat.

The 'Steel Pot' will forever be imprinted in the minds of soldiers and citizens alike around the world as an American Icon from WWII that came to the rescue and saved many a day.

USA "Steel Pots" or the M1 Helmet

Chapter VII
Dog Tags

"Well guys, now we know why they gave us dog tags. We're livin' an' dyin' in dirt an' ditches like the dogs"

A military 'dog tag' is the informal name for the identification tags worn by military personnel, named such as it bears some semblance to actual dog tags. Use of the name 'dog tags' for the military goes back to WWI; however, common use of identification tags goes back to the American Civil War (1861-1865).

The dog tag is *primarily* used for the identification of dead and wounded along with providing basic medical information (blood type) and religious preference for treatment of the military member. In some instances a brief history of inoculations is also included. Dog tags are usually fabricated from a corrosion-resistant metal or alloy; although, during wartime, they have been made with whatever metals were available. The dog tag usually contains two copies of the military member's information. The dog tag may come issued in pairs, or in singular. If issued singular, their design is for easy breakage into two separate pieces. With either design, the dog tag permits collection of personal information for notification while the other half remains with the body when battle conditions do not allow the casualty to be immediately evacuated.

In the U.S. military, two identical tags are issued, and their wearing is required at all times by soldiers in the field. One tag is worn on a long chain around the neck; the second tag is on a much smaller chain attached to the first chain. In the event the wearer is killed, the second tag is collected, while the first tag remains with the body. Some military units allow (or require) each member to wear one tag laced into their boot in lieu of the second around the neck. In the U.S. military, if the member has a medical condition that requires special attention, an additional red dog tag with the pertinent information is issued and also worn with the pair of dog tags.

Prior to the American Civil War, bodies were not often mangled; but the newer warfare technologies changed this during that war. During previous wars, those killed were usually removed from the field by comrades and identified by friends and proper records were maintained. During the American Civil War, the carnage was so massive at times, it became increasingly difficult for anyone to keep up with who had fallen. An entire company could be killed and no survivors remained for identifying the dead. Some battles were so fierce that entire units of male population between the ages of twenty and forty died in about three days. Some soldiers began pinning paper notes with their name and home address to the backs of their coats. Others stenciled the identification on their knapsacks or scratched it in the soft lead backing of their belt buckles. Because the soldiers did not want their families left wondering if they were dead or deserted, they began sewing their names into their clothing so they could be later identified.

Manufacturers of identification name badges (mainly jewelers) began recognizing a growing market as a means of emergency identification and began advertising in periodicals. Their custom badges were usually shaped to suggest the branch of service and came engraved with the soldier's name and unit. Other designs were machine-stamped tags made of brass or lead with a hole for portability with a string. They usually had an eagle or shield inscribed with patriotic phrases on one side. The other side contained the soldier's name and unit, and sometimes listed the participated battles.

In 1862, a man by the name of John Kennedy from New York wrote to the U.S. Army. He offered to furnish a type of disc for all officers and men in the U.S. Army. With his letter was a design for the disc. The U.S. National Archives has this letter along with the Army's reply (which was a summary refusal without an explanation). In the Spanish-American War during 1898-1901, soldiers continued purchasing crude stamped identification tags. Because no standardization of information was in place within the military, in many instances those tags contained insufficient to misleading information.

The first official military advocacy for issuing any form of identification tags came in 1899 during the Philippine Insurrection. U.S. Army Lieutenant Colonel Charles Pierce (Chaplain) recommended that an identity disc be included in the standard combat field kit for each combatant. With the political climate changing in Europe after the Spanish-American War, it was not many years before the clouds of huge war in Europe was looming on the world's horizon. In those threatening storm clouds, soldiers and sailors were still purchasing unofficial crude stamped identification name tags. With the magnitude of a huge European war, that could become a world war (WWI), it became evident that its carnage would far surpass that of the American Civil War. The need for personal identification on the field became an imminent necessity with the U.S. military.

The U.S. Army first authorized use of identification tags in War Department General Order Number 204, dated 20 December 1906, which essentially prescribes the John Kennedy identification tag: "An aluminum identification tag, the size of a silver half dollar and of suitable thickness, stamped with the name, rank, company, regiment, or corps of the wearer, will be worn by each officer and enlisted man of the Army whenever the field kit is worn, the tag to be suspended from the neck, underneath the clothing, by a cord or thong passed through a small hole in the tab. It is prescribed as a part of the uniform and when not worn as directed herein will be habitually kept in the possession of the owner. The tag will be issued by the Quartermaster's Department gratuitously to enlisted men and at cost price to officers..."

The U.S. Army changed its regulations on 6 July 1916, so that all soldiers were issued two tags: one to stay with the body and the other to be with the individual in charge of the burial (for record-keeping purposes). In 1918, the Army adopted and allotted the individual serial number system (service number). The soldier's name and serial numbers were then ordered stamped on the identification tags of all enlisted troops. That requirement for just the enlisted members did not last long; it also was extended to officers as well.

The U.S. Navy first officially began using dog tags when authorized by Secretary of the Navy, Josephus Daniels, in General Order No. 294 of 12 May 1917. The U.S. Navy name tags were oval, were made of Monel metal (a corrosion-resistant alloy of nickel, copper, iron, and manganese), were 1.25 by 1.5 inches, and they were perforated at one end. The single tag was designed to be worn around the neck on a Monel wire encased in a cotton sleeve. One side of the tag bore an etched print of the right index finger. The other side was stamped 'USN' and was etched with the individual's personal information. A Monel-metal chain could be used at the individual's expense.

During WWII, U.S. military dog tags could indicate only one of three religions with the designation of one letter: 'P' for Protestant, 'C' for Catholic, and 'H' for Jewish (from the word 'Hebrew'). 'NO' indicated no religious preference. Later regulations abbreviated spelling of the religious choices and went on to total spelling of them. After WWII, the U.S. Navy adopted the style of dog tags used by the Army and Air Force. Thus, a single shape and size became the standard design for all of the American military dog tags.

During WWII, there arose a recurring myth about the notch located on one end of the American dog tags. Rumors were that the notch's purpose was that, if a service member found a deceased comrade on the battlefield, one dog tag was to the be removed and taken to the commanding officer. The other dog tag was to be stuck between the teeth of the soldier. The notch supposedly was for ensuring rapid leverage between teeth when kicking the lower jaw upward toward the upper jaw. This action was to serve two purposes: (1) to ensure that the tag would remain with the body and be identified, and (2) to ensure the mouth would remain open for allowing body gases to escape for reducing bloating until it was removed or buried.

In reality, the notch did not serve either of these two purposes. In reality, the notch was designed to hold the tag in place when being imprinted in the imprinting machine used by the Medical Department during WWII. The purpose and use of the dog tag notch can be unofficially verified with the Graves Registration Service, and officially verified with the U.S. Army Mortuary Affairs. Dog tags made after WWII began losing the notch in their designs. The new imprinting machines did not require the same design for holding the tag as did the original imprinting machines.

Standard U.S. military dog tags currently are 1.25 inches wide and 2 inches long, and are issued with one 2.5mm, 24-inch long neck chain and one 2.5mm, 4-inch long duplicate chain. The secondary chain and duplicate tag are recovered if needed. The primary chain and tag remain with the soldier at all times. Dog tag ball chains were once thought to have been clipped to length for aiding captured soldiers count time in the event they were captured. The large chain was counted to 365 ball links, and the small chain was counted to 52 ball links. A soldier could then bite, or break off, one link from the large chain each day and remove one link from the small chain each week. This accounted for one year of time before the cycle consumed the chain completely.

There has never been any verified comparative study between the standard 24-inch and 4-inch chain lengths, the number of beads per length, and the typical length for 365 days and 52 weeks. This fact however, clashes with the U.S. Air Force dog tag chains. Their chains have 210 beads and are 27.5 inches in length, which can hang to the middle of some chests. A long 365 bead chain is approximately 47.25 inches, which can hang too low on the average person.

Dog tags of militaries around the world over the years have been manufactured in various shapes, sizes, colors, and materials (even fibers). Their means of suspension have also been varied ranging from twine, cord, wire, and chain. Also variant has been the information inscribed on them. In some applications the service members have used black tape as insulation material for prevented their dog tags from issuing metallic clanking sounds while in stealth modes of combat. This practice gave rise for use of rubber insulators on the dog tags, which lead to development the manufacture of dog tag insulators.

Dog tags are traditionally part of any makeshift battlefield memorial scenario soldiers create for their fallen comrades killed in action (KIA). The KIA's rifle with affixed bayonet is erected vertically atop the empty boots, with the helmet over the rifle stock. The dog tags will then usually hang from the rifle's handle or trigger guard.

Since about 1995, military dog tags have become vogue in non-military applications. Youth and adults alike have often been seen wearing custom-made dog tags. Dog tags have increasingly found their way into the youth fashion scene by way of military fashion design wishing to present a tough or militaristic image. Non-military dog tags may be inscribed with personal details, beliefs, logos, favorite quotes, photos, medical information, et cetera.

Millions of people for decades have wondered why the military identification tags are called dog tags. The truth is most likely layered beneath years of myth, legends, and actual events. Yet, beneath all of those layers are some commonality bases. So, let's look again to some events ***within*** the American

Civil War. Military dog tags very much resemble the pet identification tags in that they were made for the same purpose – *identification.* The jewelry industry saw that need and expanded new uses and designs for identification tags. By the end of the American Civil War, Union soldiers began purchasing crude animal dog tags to carry with them at all times. On these dog tags would be inscribed their personal information in the event they became casualties of war.

And thus, with the practice of those actual Union Army Soldier *dog tags*, the nickname of *'Dog Tags'* for military identification purposes was born; and it has remained common practice into these modern times.

Blank USA Dog Tags

Designer Dog Tags

Chapter VIII
Military ID Cards

"Where's that wise guy that said he lost his ID card and so couldn't be joinin' us today?"

At the time of writing this book, the modern world takes fingerprinting for granted. People everywhere think nothing of having to submit for being fingerprinted. The concept for being fingerprinted, though, is surprisingly quite old; whereas the science of it is relatively young in time standards. In Nova Scotia, picture writings were discovered dating back hundreds of years; they were of a hand with ridge patterns. In ancient Babylon, fingerprints were discovered for use on clay tablets in business transactions. In ancient China, thumb prints have been found on clay seals. Thus, the concept for fingerprinting as a means of personal identification is not new to the modern world – its concept existed in antiquity.

It was actually in 14th century Persia when a discovery was made that fingerprints very unique. It was there that various official government documents have been found with impressions of fingerprints. One government official (a doctor) had even observed that no two fingerprints imprinted were exactly alike.

In July of 1858, Sir William James Herschel, Chief Magistrate of the Hooghly District in Jungipoor, India, first began using fingerprints on native contracts. With one transaction Mr. Herschel even had a local businessman (Rajyadhar Konai) impress his handprint on their contract. The practice of fingerprinting on contract documents revealed some surprising discoveries. For the natives, personal contact with the document was believed to become more binding than if they simply signed the document. Thus, the first wide-scale use of fingerprints was set in motion, not based on scientific evidence, but on superstitious beliefs.

As Mr. Herschel's fingerprint collection grew, he began noticing the inked impressions could, indeed, prove or disprove individual identities. Even though his experience with fingerprinting was very limited, it was Mr. Herschel's private conviction that all fingerprints were unique to the individual, as well as permanent throughout an individual's life. That personal conviction inspired him to expand the use of fingerprinting. Afterwards, the English began the practice of using fingerprints on a very wide scale.

In 1882, Gilbert Thompson of the U.S. Geological Survey in New Mexico used his own thumb print on a receipt document for the express purpose of preventing its forgery. This became the first known use of 'fingerprinting' in America.

In 1883, with the book, *Life on the Mississippi* by Mark Twain, a murderer was identified by the use of fingerprint identification. Later with another book, *Pudd'n Head Wilson* also by Mark Twain, he writes of a dramatic court trial involving fingerprint identification.

In 1888, Sir Francis Galton, a British anthropologist (and cousin of Charles Darwin), began observations of using fingerprints as a definite means of individual identification. Then in 1892, he published a book titled *Fingerprints*, which established the individuality and permanence of fingerprints for everyone. He was able to scientifically prove that fingerprints never change over the course of an individual's lifetime, and that no two fingerprints are exactly the same. According to his calculations, the odds of two individual fingerprints being the same were 1 in 64 billion. Mr. Galton identified the special characteristics by which fingerprints can be identified, and his book also included the first classification system for fingerprints.

In 1900, the UK Home Secretary Office conducted an inquiry into 'Identification of Criminals by Measurement and Fingerprints.' Mr. Edward Richard Henry (later Sir Edward R. Henry) appeared before the inquiry committee to explain the system published in his recent book *The Classification and Use of Fingerprints*. The inquiry committee recommended the adoption of fingerprinting as a replacement for the relatively inaccurate system in place at the time. In July of the next year, the Fingerprint Branch at New Scotland Yard was created using the Henry System of Fingerprint Classification.

In 1902, the first systematic use of fingerprints in the U.S. was set in place by the New York Civil Service Commission for testing. Then in 1903, the New York State Prison System began the first systematic use of fingerprints in the U.S. for identifying criminals.

In 1904, the U.S. Government began considering the use of issuing identification cards for its military members. From those considerations, the Department of War (DOW) approved and instituted Form Number Two, which became known as the U.S. Military ID Card. It contained the military bearer's name, rank, service number, photo, signature, and thumbprint. It became a definite means for immediate secure identification, as well as being portable enough for transporting in a wallet.

In 1905, the U.S. Army began using fingerprints when issuing Military ID Cards for its members. In 1907, the U.S. Navy began using fingerprints when issuing Military ID Cards for its members. In 1908, the U.S. Marine Corps began using fingerprints when issuing Military ID Cards for its members.

After the third Geneva Convention, the name for the U.S. Military ID Card was changed to the United States Uniformed Services Privilege and Identification Card (USPIC). However, that name never quite caught on and the simplistic older name of 'ID card' remained favored among the service members. A few years later, Form 1173 was instituted for identifying dependants of military service members. This came about because while service members were deployed, their dependants needed access to base medical and commissary facilities. Therefore, they needed some form of identification as well.

In 1949, the U.S. Department of War was changed to the Department of Defense (DOD). Many of the forms the War Department had been using were changed to a 'DD' classification. Thus, the military ID Cards became the current DD Form 2 and DD Form 1173. The Military ID Card varies in color, which indicates the holder's status as follows:

- Green for Active Duty and Active Duty Reserve Components.
- Red for Reserve Components under age sixty and dependents of Reserve Components.
- Tan for dependents of Active Duty and Retired Components.
- Blue for Retirees.

As previously started, the USPIC (Military ID Card) is actually an identification document issued by the U.S. military for the purpose of identifying an individual as a member (or their dependent) of the Armed Forces. In addition to the Military ID Card, the U.S. military issues Military ID Cards to its Military Civilians. The form numbers vary from service to service as they are service specific. Grey is the color for Active Civilian Duty Status and blue is the color for Civilian Retiree Status.

All the before mentioned information on Military ID Cards was basically true until 2003.

In 2003, DD Forms 2 and 1173 were retired and replaced by the U.S. military with Common Access Cards (CAC). The CAC is a 'smart card' that was issued as standard identification for active-duty military personnel, reserve personnel, civilian employees, and eligible contractor support personnel. The CAC is also utilized by other non-DOD government agencies for their employees.

The CAC is used as a general identification card as well as an authentication tool for enabling access to agency computers, networks, and certain facilities. For the DOD, the CAC also serves as an identification card under the Geneva Convention. Because the CAC is a smart card, it enables encrypting and cryptographically signing email, and facilitates electronic authentication. The CAC establishes a definite *authoritative process* for the use of identity credentials. Electronic and manufacturing MilSpecs were used in the development of CAC technology. With CAC technology, the military migrated its identification needs from simple ID cards to an *identity credentialing process* into the world of tomorrow.

Even though this entry on Military ID Cards has provided a very brief overview of that topic, just guess when, where, and through whom, the apparent <u>*first*</u> identification cards were implemented? It was in the early 1800s in France, and by French General Napoleon Bonaparte!

Prior to 1800, the French Republic had created a degree of freedom unheard of in Europe. This allowed free speech and gave workers the right to change their employment and relocate anywhere. In contrast, in most of Europe at this time, including Britain, the population's majorities lived in various forms of oppression, such as indentured servitude. Unfortunately, though, in France, their free market and mobility of labor were driving up wages.

General Napoleon sought a transformation of the free French Republic society into an empire which was one of a tightly controlled police state. He envisioned a concept of issuing a national identity card, and it was the first step in his transformation plan. Thus, in 1803, General Napoleon's Police Chief reinstated the Livret ID Card and updated it with new identity features. The Livret ID Card was the Worker's Passbook was used by the Old French Regime.

To obtain employment, workers were required to surrender their Livret ID Card to their employer. When taking new employment, workers had to obtain their Livret ID Card from their employers. This action, though, required obtaining their employer's permission to leave. The Livret ID Card also acted somewhat as an Internal Passport, making it very difficult for relocating to seek better work or better wages. Moving from one town to another required a set of visas with the Livret ID Card. Without these, employment was illegal.

Napoleon's national ID card scheme failed to be completely successful for two reasons: (1) there was a labor shortage (as a result of the war) which made employers willing to take on workers without a card, and (2) the workers had self-help networking groups who helped their fellows find lodgings and employment. They helped each other get around the government system. The authorities tried to introduce state schemes for displacing these underground networking groups, but they were only partially successful. After the retreat of the French Empire, and the fall of General Napoleon, fragments of the census and control systems he had introduced were retained because they were deemed too useful and efficient for abolishing.

Thus, General Napoleon's concept of issuing a national identity card was actually the original ancestor of all modern ID card systems.

Of all the configurations with 'body life lines' and their shapes, *only one select special* set of them combine for making fingerprint patterns ... ***a truly unique personal imprint***. Therefore, make a lasting impression for others to follow.

A 'sanitized' U.S. Navy ID Card from the 1920s

A 'sanitized' U.S. Armed Forces
ID Card from the 1970s

CHAPTER IX
THE P-38

"Awww Bob, cannot believe this! Dropped my P-38!" "Got it, Chief! Now let's get off the beach before we get our butts shot off!"

Every older military member knows what a standard issue P-38 is. It isn't the P-38 Fighter Aircraft, and it isn't the P-38 Automatic Pistol. It is that nifty little P-38 Can Opener. Many in the military wore one along with their dog tags on their neck chains. Some trivia about that small lightweight P-38 Collapsible Can Opener is rather interesting.

It was developed during WWII, and reportedly was the results of a thirty-day rapid design project by the U.S. Army in the summer of 1942. The origin of its name is unclear. Some claims are that it required exactly thirty-eight punctures around the rim of a C-Ration can for successfully opening the can. Other claims state it performed with the speed of the twin-engine Lockheed P-38 Lightning Fighter Plane. Whatever the origin for the P-38 Can Opener (and its name), this little device has been considered one of the most perfect small inventions ever designed for use in combat.

The P-38 Can Opener also became known as a 'John Wayne' by the U.S Marine Corps. This famous labeling was either because of the tool's tremendous toughness and dependability, or was from a possible legend; that of an unsubstantiated story about the actor actually being seen opening a can of rations while in an unidentified training film. Knowing the dependably of both the P-38 Can Opener, and the love of actor John Wayne for the USA and the military, the labeling was most likely from both sources. Hoorah for the Marine Corps!

In the modern militaries, C-Rations in cans have largely been replaced by MREs (Meals Ready to Eat). MREs come in a pouch and all that needs being added is water. Many still prefer the C-Ration cans because the MRE pouches are more vulnerable to puncture damage than the cans. However, if one does not have a P-38, one then must rely on a knife or bayonet for opening the C-Ration can. Those two devices used for opening cans are significantly slower, however, than the P-38.

MilSpecs were used by the military and industry as bid, manufacturing, and management standards as well as guidelines for developing the P-38 Can Opener. Because of MilSpecs, soldiers, sailors, and airmen in the outlands could eat their rations when time was of the limited essence. Because everyone in the development and procurement process was operating from the 'same page' (MilSpecs), their efforts helped to sustain soldiers for winning a war.

In the military of days gone by, many were seen wearing crosses and P-38s along with their dog tags. They wanted the shield of protection the Cross of Christ symbolized. They also wanted always being with a means of opening a can of food. There is seemingly a spiritual parallel with this practice: The Message of the Cross provides The Key to rapid recovery; and with lightning speed, the Word of God can pierce through any barrier. Barriers can come in many forms – our own and from the enemy. For whatever reason, God's messages can rapidly get to the bottom of any issue.

"For the word of God is living and powerful, and sharper than any two-edged sword, piercing even to the division of soul and spirit, and of joints and marrow, and is a discerner of the thoughts and intents of the heart" (Hebrews 4:12).

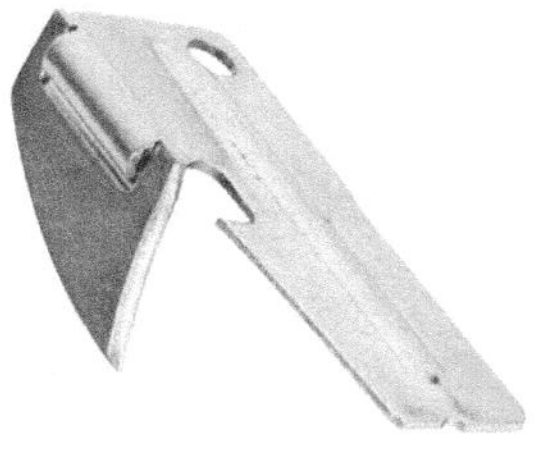

U.S military P-38 Rations Can Opener

CHAPTER X
THE MAE WEST

"Ya know Sarge, I'd feel a lot better about all this if I was wear'n one'a them Mae Wests!"

The 'Mae West' was the first inflatable life preserver used by the military on a large scale. It was invented in the late 1920s, and its nickname originated because someone wearing the inflated life preserver often appeared to be as physically endowed as the actress Mae West. In March 1926, Mae West's play *Sex* was in rehearsal. Prior to its Broadway run, the play had out-of-town try-outs in Connecticut.

Sailors on shore leave bought up the tickets and became a most appreciative audience for those previews. Because of this, Mae West used to say 'I'll never forget my debt to the Navy.' With her hourglass figure, and sinful stroll on top her five-inch high-heels, she was a natural bombshell on the stage. From that commonly uttered phrase of debt to the U.S. Navy, Allied sailors and soldiers nicknamed their inflatable life jackets 'Mae West' in honor of her famously buxom-figured actress.

Life preservers and life jackets; we take them for granted when going boating or on cruises. They are an 'expected' piece of life preserving equipment for maritime ventures in this era. However, it was not so long ago that personal flotation devices were not as common as they now are. Most examples of primitive life jackets from antiquity can be traced back to inflated bladders of animal skins or hollowed out and sealed gourds. Their use, though, was primarily used for support when crossing the deeper zones of streams and rivers.

Surprisingly, personal flotation devices were not part of the survival equipment issued for naval sailors up to the early part of the 19th century. It wasn't until lifesaving rescue services were being formed, that the personal safety of rescue boat crews was addressed. Their heading out to sea for pulling back boats (and usually in horrendous sea conditions) generated the need for personal flotation devices. Earliest designs have been recorded as just simple blocks of wood or cork pads.

Credit, though, for the modern life jacket is somewhat controversial. Some historical records credit the invention to a certain Captain Ward, who was a Royal National Lifeboat Institution inspector in the United Kingdom. Other records credit the invention to Mr. Peter Markus (U.S. Patent 1694714). Still other records credit the invention to James Boyle before WWII, who was founder of the Air Cruisers Company. With an invention controversy over personal flotation devices having been unraveled for decades, exploration into some workings of the inventors would be in order.

In 1854, Captain Ward created a cork vest to be worn by lifeboat crews for both weather protection and buoyancy. The rigid cork material eventually came to be augmented with pouches containing watertight cells filled with kapok (a vegetal material). These soft cells were more flexible and comfortable to wear in comparison with devices utilizing hard cork. Kapok was being used for buoyancy with many navies by this time. Foam eventually replaced kapok for inherent buoyant flotation qualities.

In 1935, James Boyle founded the Air Cruisers Company (now a part of Zodiac Group). Prior to World War II, he invented the inflatable life jacket which was called the 'Mae West.' Since then, most every significant development in the field of aviation inflatable safety technology has come from the Air Cruisers Company. Air Cruisers produced the first one-person life raft, the first inflatable life vests, the first vacuum packed inflatable life vest, the first inflatable aircraft evacuation slides and slide rafts, and the first helicopter fixed floats and the first helicopter pop-out inflatable floats.

On 16 May 2005, *The New York Times* published an article when California inventor Andrew Toti died in March at the age of 89. Mr. Toti was memorialized in newspapers from coast to coast as the man who'd invented the Mae West life preserver which saved countless lives of servicemen in WW II. According to *The New York Times*, in a letter to *The New York Times*, Alvin A. Markus of Los Angeles claimed it was his father, Peter Markus, and not Andrew Toti who was the inventor of record. Mr. Markus included photocopied patent documents that dated from the late 1920s and newspaper articles from the 1930s and 40s.

The documents show that Mr. Markus won his first patent for an 'inflatable life preserver' in 1928, and two patents for improvements in 1930 and 1931. An article from *The Tulsa Daily World* in 1936 tells how the life jacket Mr. Markus developed saved the lives of 98 of 100 men when the Navy dirigible *Macon* crashed in the Pacific in 1935. The term 'Mae West' was not used however because it had not yet become commonplace.

There are other claims and later patents for similar flotation devices – even one invented by a Frank Manson of Ohio in 1935. How many others laying claim for originating the inflatable life vest is most difficult in knowing. However, according to Rothwell, Figg, Ernst, and Manbeck (a law firm in Washington, DC), it does not change the fact that

Mr. Peter Markus did, indeed, *get there first with an invention patent*. A patent search by their office at the request of *The Times* found that Peter Markus's 1928 claim is the earliest U.S. Patent for an inflatable life vest 'similar' as the one commonly known as the Mae West.

From a 'who got there first?' perspective, the law firm seems to have centered on the long time controversial target. However, had all parties involved been operating from the same page of MilSpecs, all this controversy would most likely never have gotten started to begin with. Operating from the same page of MilSpecs would have maintained some relevancy and order in what has come to appear a bit of brainstorming invention chaos. Looking past all the flotation clutter, a very pressing need was identified, models developed, types successfully fielded, and many thousands of lives were saved.

Regardless of the *prime credit*, all of the actors seemed to have obtained a portion of the *end credits*. For those enduring invention efforts (and without much 'territorial' quarreling either) multitudes of lives are forever most thankful.

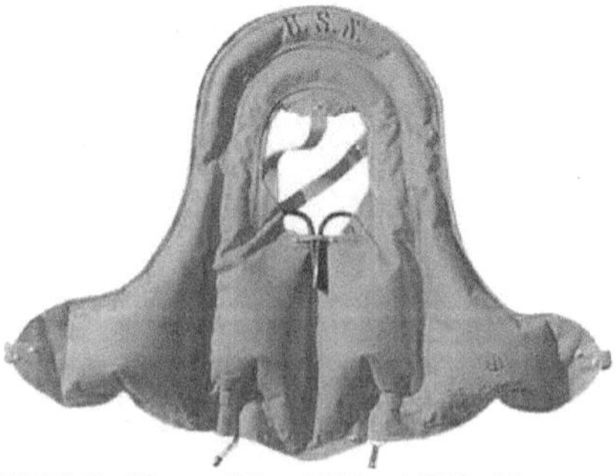

WWII U.S. Navy Mae West Life Preserver

CHAPTER XI
JEEP WORKS

"Yeah well Sarge, the Jeep works great, but that new gun operator...well...think he's gonna need a few adjustments here and there"

Over the decades, there have arisen many explanations for the origin of the word 'Jeep' and all of them have proven difficult in verifying. The years have seemingly held mostly to the theory of a military designation – GP. GP apparently began the term of Jeep and still holds to that vehicle name. The designation GP was for Government Purpose or General Purpose and became phonetically slurred into the word 'jeep.'

Another view was launched by Mr. Lee Ermey on his television series *Mail Call.* The view disputes this explanation claiming the vehicle was designed for specific duties, was never referred to as General Purpose, and that it was highly unlikely that the average Jeep-driving GI would have been familiar with this designation. Another view is the Ford GPW abbreviation actually meant 'G' for Government Use; 'P' to designate its 80-inch wheelbase; and 'W' to indicate its Willys-Overland designed engine. With all these facts, there certainly seems to be somewhat of a reasoning disconnect, though, for *phonetically* one cannot derive the term 'jeep' from GPW.

Other views also suggest that soldiers during the early Jeep era were so impressed with the new vehicles that they informally named it after Eugene the Jeep, a character in the Popeye cartoons created by E. C. Segar. Eugene the Jeep was Popeye's 'jungle pet' and was small, able to rapidly move, and solve seemingly impossible problems.

The dictionary of military slang, *Words of the Fighting Forces*, by Clinton A. Sanders, published in 1942, provides this definition for the term of Jeep: 'A four-wheel drive vehicle of one-half- to one-and-one-half-ton capacity for reconnaissance or other army duty. A term applied to the bantam-cars, and occasionally to other motor vehicles (U.S.A.) in the Air Corps, the Link Trainer; in the armored forces, the one-half ton command vehicle. Also referred to as any small plane, helicopter, or gadget.'

Interestingly, this definition is also verified by the usage of the term 'jeep carrier' in reference to the U.S Navy's small escort carriers.

Early in the year 1941, Willys-Overland Motors demonstrated the Jeep vehicle's durable off-road capability by having it drive up the steps of the United States Capitol, driven by Willys test driver Irving 'Red' Haussman. Mr. Hassman had recently overheard soldiers at Fort Holabird calling the vehicle a jeep. When asked by columnist Katherine Hillyer of the *Washington Daily News* what it was called, Mr. Haussman answered, 'It's a jeep.' The original trademark brand-name application of 'Jeep' was filed in February 1943 by Willys-Overland Motors. That same name is also used in generic terms in lowercase (jeep) for vehicles inspired by the Jeep which are suitable for navigation on rough terrains.

As for the actual beginnings of the Jeep, its origin came when it was obvious that America was eventually going to become involved in the war raging in Europe. On 7 July 1940, the U.S. Army Quartermaster Corps contacted 135 companies asking for working prototypes of a four-wheel-drive reconnaissance car. The U.S. Army List of Specifications called for the following deliverable items in the new four-wheel-drive reconnaissance car:

1. A driving front axle with 2-speed transfer case including provisions for disengaging the front axle drive.
2. A body of rectangular design with folding windshield and 3 bucket seats.
3. Increased engine power (presumably in respect to the Belly-flopper prototype).
4. Means for towing.
5. 30-caliber machine gun mount.

6. Blackout lighting.
7. Oil-bath air cleaner.
8. Hydraulic brakes.
9. Full floating axles.
10. Wheelbase of 80".
11. Maximum height of 40".
12. Maximum weight of 1275 lbs.
13. Approach and departure angles of 45 and 40 degrees, respectively.
14. Must reach 50 mph on hard surface.
15. Special bracing for a pintle hook setup.
16. No aluminum to be used for cylinder head.
17. At least four cylinders.
18. 8 of the 70 vehicles to be made had to be four-wheel-steering.

The Army established what seemed like an impossible deadline of forty-nine days for furnishing the working prototype to the War Department (now the Department of Defense). Only two companies responded to the request, The American Bantam Car Company and Willys-Overland Motors. Willys-Overland requested an extension but it was denied. Bantam Cars, in spite of their own problems, completely laid out plans for their prototype in two days, and the next day estimated the total cost of the vehicle.

On 22 July, Bantam's bid was submitted, complete with blueprints. Much of their vehicle required assembly from existing off-the-shelf automotive parts. Their hand-built prototype was completed in Butler, Pennsylvania, and driven to Camp Holabird, Maryland, for testing by the Army on 21 September 1940. The vehicle met the Army's specific vehicle criteria, but its engine could not meet the Army's torque requirements.

The U.S. Army was of the opinion that Bantam Cars were too small for supplying the required number of vehicles needed. So, the Army supplied the Bantam design to Willys-Overland Motors and Ford Motor Company and were encouraged to make their own changes and modifications. This resulted in the Ford 'Pygmy' and the Willys 'Quad' prototypes which looked very similar to the Bantam BRC (Bantam Reconnaissance Car) prototype. Fifteen hundred each of these three models were built and extensively field-tested.

The War Department required large numbers of these vehicles to be manufactured in a relatively short time. Therefore, Willys-Overland Motors granted the U.S. Government rights for licensing another company to manufacture vehicles using their specifications. The Army chose Ford as a second supplier for building the Jeeps to the Willys' design. Willys supplied Ford with a complete set of plans and specifications. American Bantam (creators of the first Jeep) spent the rest of the war building heavy-duty trailers for the Army.

Jeeps were not only used by the infantry and artillery, but ended up becoming used by every facet of the U.S. military. Jeeps found many other purposes including cable laying, sawing, machining, firefighting, field ambulances, utility vehicles, tractors, and with suitable wheels it could even run on railway tracks. As part of the war effort, nearly thirty percent of all Jeep production was supplied to Great Britain and to the Soviet Red Army during WWII.

After WWII, manufacturing of the Jeep over the years changed from Willys to Kaiser-Jeep, American Motors Corporation (AMC), Chrysler, and the subsequent Daimler-Chrysler Groups. In the U.S. military, the Jeep became replaced by a number of vehicles, of which the Ford M151 MUTT was one. The M151 became the forerunner of the HUMVEE (or Hummer as it is commonly known).

The Jeep has been widely imitated around the world, and has probably become *the most widely imitated motor vehicle of all time*. The Museum of Modern Art described the Jeep as a masterpiece of functionalist design. Combat Journalist Ernie Pyle called the Jeep, along with the Coleman GI Pocket Stove, 'the two most important pieces of noncombat equipment ever developed.' During the Jeep's service in WWII and especially Korea, its name was referred to as '***J***ust ***E***nough ***E***ssential ***P***arts' by the GIs due to its very basic design. And thus, another version for the name of Jeep was born, applied, stayed, and has become commonplace for the entire world to evaluate.

One very serious consideration beyond the Jeep's name is with the use of MilSpecs. Had MilSpecs NOT been used as a *common center of reference* for all the players in management and manufacturing, the Jeep would NEVER have entered WWII as fast as it did. Had the Jeep NOT entered the war as fast as it did, the outcome of WWII might have been somewhat different. However, from those scant first few prototypes, to the first 3700 units produced in 1941, millions of military and civilian Jeeps have been manufactured, leaving very prolific pages within the manuscript we sometimes call...*'the rest is history.'*

1940 Bantam Reconnaissance Car
The First Jeep (both photos)

1941 Willys Jeep

WWII Armored Jeep

2011 Jeep Rubicon

CHAPTER XII
M&Ms

"Sure hope the chow supply guys got some'a them new M&Ms on this God forsaken island"

In 1954, the popular marketing tagline and commercial of 'Melts in your mouth, not in your hands' began for the M&M Candy. That year the M&M Peanut Candy was also introduced. As most everyone knows, M&Ms are those colorful button-shaped candies which usually are packaged in small shinny bags. Their shells each have the lower-case letter 'm' imprinted on one side. Their fillings come in a variety of flavors: milk chocolate, dark chocolate, peanut butter, crisped rice, mint chocolate, peanuts, almonds, orange chocolate, coconut, pretzel, and wild cherry.

Sold in over one hundred countries around the world, a somewhat forgotten and little known fact is how M&Ms were originated. M&Ms are produced by Mars, Incorporated in the USA, and are named after the surnames of the company's founders, the Mars and Murrie brothers. Forrest Mars, Sr., the founder of the Mars Company, came up with the idea in the 1930s during the Spanish Civil War. It was then he observed soldiers eating chocolate candy with a hard shell of tempered chocolate surrounding the outside in an effort to prevent the candy from melting. On 3 March 1941, Mr. Mars received a patent for his own unique and improved process of candy coating. Later that year, production of the candies began in a factory located at 285 Badger Avenue in Clinton Hill, Newark, New Jersey.

M&M candy production was in partnership with Bruce Murrie, the son of Hershey's Chocolate president, William Murrie. This arrangement allowed the candies to be made with Hershey chocolate (who then had control of war rationed chocolate). The first M&Ms, as a war deliverable to soldiers, were packaged in cardboard tubes, and were made in five colors: red, yellow, brown, green, and violet. Thus, in 1941, American servicemen serving in WWII were introduced to a field stable combat ration candy; one that would become America's favorite candy coated chocolate.

During WWII, the M&M candies were exclusively sold to the military. The practicality of the candies during WWII generated an increase in production requiring a larger manufacturing capacity. Its factory relocated to larger facilities at 200 North 12th Street in Newark, New Jersey, where they remained until 1958 when it was relocated again to a larger facility in Hackettstown, New Jersey.

In 1948, the packaging was changed from cardboard tubes to the commonly known small shinny bags of today. In 1950, the very first 'M' was imprinted in black. It gave them their unique trademark and was eventually changed to white. The imprinting of 'M' was for Mr. Forrest Mars Sr., and the other one was for Mr. Bruce Murrie. During this same time, the world famous M&Ms characters made their television debut. M&M's were finally introduced internationally in 1980 and have been an American classic and favorite ever since.

Large numbers of vendors and MilSpecs apparently did not play an enormous role in the development with one of America's favorite candies. There are times when a serious need arises and there happens to be someone already working on the answer. Quite often during those times the manufacturer is usually aware of the required testing and the MilSpecs supporting the needed processes. It is then the manufacturing process and the MilSpecs blend in a marriage for 'saving the day' and winning wars.

From careful observance, to thoughtful contemplation, to researching new manufacturing applications, blessed with partnerships, a major success was born. The military obtained it first and the entire world began enjoying it a few years later.

MilSpecs, though, were part of the conduit for its achievement. Who would have ever thought those delicious little button shaped candies called M&Ms were first made for the U.S. military? *M&Ms, melts in your mouth, and not in your hands...*

Front of an Early M&M Candy Dispenser

Photo Depicting Changes in M&M Candy

CHAPTER XIII
GATLING GUNS

"Hey Sarge, ya know, think them last few thousand rounds finally got'em this time"

The Gatling Gun was the first of rapid-firing guns and was designed by the American inventor Dr. Richard J. Gatling in 1861. Its operational design centered on a recurring (or cyclic) multi-barrel concept. Its design offered the battlefield warrior a weapon with a rapid and continuous rate of fire – without manually reloading through opening its breech. The original gun design was intended for very limited use as a field weapon.

With the Gatling Gun's cyclic multi-barrel concept, each barrel fired a single shot when attaining a specific point in the cycle. The spent cartridge was ejected and a new round was loaded. Because the gun had six barrels, during this cyclic process the barrel cooled down before its next firing. This novel recurring design configuration permitted achieving higher rates of fire without the barrels overheating. Its six rotating barrels were turned by a hand crank and fired loose cartridge ammunition from a gravity fed ammo hopper system.

Effectiveness of the early Gatling Guns relied mainly on the operators. Unskilled operators could achieve relatively high rates of fire of around 200 rounds per minute (rpm). Skilled operators could achieve higher rates of fire of around 400 rpm. By the mid 1880s, design improvements boosted the firing rates even higher to around 1,200 rpm. Also by the mid 1880s, the Gatling Gun was being exported to countries throughout Europe and Asia.

Even though its first combat use was by Union Army Forces during the American Civil War in the 1860s, it was not fully accepted by the U.S. Army until 1866. Later it proved being a tremendous asset for the Army with the assault on San Juan Hill during the Spanish-American War. During that assault in July 1898, there was a four-gun battery of ten-barrel Gatling Guns (in .30 caliber ammunition) which was formed into a separate detachment led by First Lieutenant John 'Gatling Gun' Parker. Three of the four Gatling Guns where equipped with swivel mounts. During the American charge up the San Juan and Kettle Hills, those three swivel-mounted Gatling Guns fired approximately 18,000 rounds in eight and one-half minutes against Spanish troop positions along the crest of both hills.

The bulleting result was massive carnage.

Early in the 20th century, the Gatling Gun application gave way to another invention. This one employed recoil (gas-operated) non-rotating machine guns and automatic cannons, instead of using multiple rotating barrels for rapid firing, and became known as the 'machine gun.' Because of the machine gun's portability, the Gatling Gun concept fell into disuse for several decades, only to reappear immediately after WWII.

At the end of WWII, the U.S. Army began considering new directions for future military aircraft gun designs. With the newly developed jet aircraft by the Germans during WWII, the higher speeds of jet-engine fighter aircraft indicated that achieving an effective number of hits would be very difficult without a higher rate of fire.

In response to this field Operational Requirement (OR), the military resurrected the concept of the Gatling Gun. The original Gatling gun had fallen behind in developmental favor because of two main reasons: (1) portable machine gun usage, and (2) the need for an external power source for rotating the barrel assembly. However, with the newer future generation of jet-powered fighters on the horizon, those sleek jet aircraft offered sufficient power for operating an electric powered Gatling Gun. The concept of powering a Gatling Gun with an external electric power source was actually not a new one at the end of WWII. Dr.

Richard Gatling himself had conceived that idea and filed a patent on it in 1893.

In 1946, the U.S. Army issued General Electric (GE) a defense contract for Project Vulcan. This contract was to develop an electric powered six-barrel weapon capable of firing 7,200 rpm. The first GE Vulcan Gatling Gun prototypes were .60 calibers, were ground-fired in 1949, and achieved 2,500 rpm. By 1950, the Vulcan's rate of fire was increased to 4,000 rpm. By the early 1950s, the U.S. Air Force suggested high velocity alone might not be sufficient to ensure target destruction and tested 20mm and 27mm alternatives in 1952. Eventually, the 20mm cartridge was determined to have the best desired balance of projectile trajectory, explosive weight, and muzzle velocity. With more testing and improvements, by 1954, the T-171 Vulcan Gatling Gun was produced that could fire 6,000 rpm.

The first aircraft the T-171 Vulcan Gatling Gun became installed in was the prototype Lockheed F-104 Starfighter. That first prototype was completed in early 1954 and first flew on 4 March at Edwards AFB. Test flights revealed that the T-171 Vulcan suffered problems with its linked ammunition, was prone to misfeed, and presented a foreign object damage hazard with discarded links. A linkless ammunition feed system was developed for the T-171 Vulcan and its modifications warranted a weapons systems upgrade. Afterwards, the T-171 Vulcan was redesignated as the M61 Vulcan Gatling Gun. The M-61 Vulcan Gatling Gun, and its variants, has subsequently become the standard armament cannon of many U.S. fighter aircraft.

Since WWII, Vulcan Gatling Guns have varied in design from electric power operation to being hydraulically and pneumatically driven. Their rates of fire vary from around 3,000 to 7,200 rpm. Their calibers range from around 7.62mm to 30mm. ***And, they are extremely destructive and lethal***. Thus, the Gatling Gun actually ended up becoming the first weapon of mass destruction for the regular soldier. Interesting, though, documentation supports why Dr. Richard Gatling originally invented his gun in the first place; he had written of:

- First creating it to reduce the size of armies;
- Secondly, to reduce the number of deaths by combat and disease;
- Thirdly, to show how futile war was.

Strange isn't it – how some illustrations can become twisted into a far worse nightmare than originally intended?

1876 Model Gatling Gun

M-61 Vulcan Gatling Gun

Gatling Guns installed on an AC-130 Gunship

CHAPTER XIV
RATIONS

"Ya know, ain't so crazy 'bout eatin' outta cans, but after all this mess is over, I'm never gonna complain about our chow ever again"

The 18th century French military leader Napoleon Bonaparte is quoted as saying, 'An army marches on its stomach.' Recorded history illustrates this statement is very factual; armies may be mighty, but they can only be as strong as their kitchens. Food and water rations were, and are, *definite life requirements* to be supplied for warriors maintaining conflicts in any location.

With the introduction of baking processed cereal grains, a more reliable source of stored food was obtained. Ancient Egyptian sailors carried a flat brittle loaf of millet bread called the Dhourra Cake. Roman soldiers were supplied with Buccellum Biscuits. When King Richard The First of England left for the Third Crusade (1189 to 1192), his troops carried Muslin Biskits (made from a mixture grain of barley, rye, and bean flour).

The ancient bakers made biscuits as hard as possible so they would be durable in transit. Their hardness made them resistant to humidity and other weather elements. Because the biscuits were so hard and dry, their characteristics gave them the name of 'Hardtack.' To soften hardtack biscuits, it was often dunked in salt water, coffee, other liquids, or cooked in a skillet meal. Baked hard and stored properly, hardtack could stay intact for years if it was kept dry. In preparation for long voyages, hardtack was baked four times, instead of the usual twice. It was also prepared six months before sailing or deployment.

From the American Revolutionary War to the Spanish-American War, the general food ration for the U. S. Army was the Garrison Ration. It consisted of salt meat, hardtack, and vegetables (when possible). There was also a spirit ration. In 1785, it was set at four ounces of rum; then, in 1790, it was reduced to two ounces of whiskey, brandy, or rum. In 1832, the spirit ration was discontinued and replaced with a ration of coffee and sugar, which was increased in 1836. In 1846, the spirit ration was reinstated for issue to troops engaged in construction or surveying duties; this ration was discontinued in 1865.

After the Spanish-American War, newer developments in food rations were implemented and ended up being put into place for WWI. Three types of rations came into usage by the American forces: the Iron Ration, the Trench Ration, and the Reserve Ration.

The Iron Ration was in use from 1907 to 1922, and was the first attempt at providing an individual ration for issue to soldiers in the field. The ration kit consisted of three, three-ounce cakes made from a beef bouillon powder and parched cooked wheat, three one-ounce bars of sweetened chocolate, and packets of salt and pepper. The ration was issued in a sealed tin packet weighing one pound. It was designed for emergency use when the troops were unable to be supplied with food.

The Trench Ration was in use from 1917 to 1918, and was issued in the early part of WWI for addressing a specific problem. Soldiers fighting in the front lines needed daily supplies of rations. However, rations cooked and prepared in field kitchens were sometimes spoiled from enemy gas attacks. Thus, the Trench Ration became the combat solution. This ration kit contained a variety of canned meats (salmon, corned beef, sardines) which were commercially procured and sealed in large tin boxes covered in canvas. The boxes were bulky and heavy, and soldiers often began tiring of the limited menu.

The Reserve Ration was used from 1917 to 1937, and was first issued during the latter part of WWI to feed troops who were away from a garrison or field kitchen. It originally consisted of twelve ounces of fresh bacon or one pound of canned meat (which was usually corned beef). Additionally, two eight-ounce cans of hard bread (or hardtack biscuits), a packet of 1.16 ounces of pre-ground coffee, a packet of 2.4 ounces of granulated sugar, and a packet of 0.16 ounces of salt were issued. A separate Tobacco Ration of 0.4 ounces of tobacco and ten cigarette rolling papers was also included. The Tobacco Ration later contained brand-name machine-rolled cigarettes.

After 1918, the U.S. Army food ration system went through several revisions, and eventually led to the adoption of the following ration categories:

- A-Ration; a Garrison Ration and was the most valued of all rations. It contained fresh, refrigerated, or frozen foods, and was prepared in dining halls or field kitchens.
- B-Ration; a Field Ration which contained canned, packaged, or preserved foods. It was normally prepared in field kitchens without refrigeration.
- C-Ration; an Individual Ration which was completely pre-cooked, and was a ready-to-eat canned individual meal.
- K-Ration; an Individual Ration which was designed as a short duration individual assault ration for paratroopers and other specialized light infantry forces. It was discontinued in 1948.
- D-Ration; an Emergency Ration Bar of concentrated chocolate combined with other ingredients to provide high calorie content.

On into WWII, there were various other special rations developed for specific circumstances, which included Type X Rations, 5-in-1 Rations, 10-in-1 Rations, Mountain Rations, Jungle Rations, The Assault Lunch, The Assault Ration, The Aircrew

Lunch, The AAF Combat Lunch, Parachute Emergency Ration, Life Raft Rations, and Airborne Lifeboat Rations.

The K- and D-rations were discontinued after WWII, but canned wet rations in the form of the C-Ration continued until 1983, when they were replaced by the Meal, Ready-to-Eat (MRE). In 1958, the C-Ration name was changed to Meals, Combat Individual (MCI). Even though MCI was the official designation, soldiers still preferred calling them C-Rations until they were discontinued.

The MRE (as is commonly known) is a self-contained (reportedly) individual field ration in lightweight packaging procured by the U.S. military for their service members for use in combat (or for when other field conditions arise where organized food facilities are not available). Stored within the plastic retort pouches are the dehydrated food which makes up the MREs. The standard issue of MREs in 1986 was limited to a menu of twelve entrees but, as of the writing of this book, has been expanded to twenty-four entrees. In 1992, a Flameless Ration Heater (FRH) was included in the MRE kit. The FRH is a water-activated exothermic reaction product which emits heat for providing service members in the field with hot meals.

As of the writing of this book, the following military food rations for field use are available to U.S. troops:

- A-Ration; Fresh food prepared on-site (or nearby and transported).
- B-Ration; A unit-sized packaged and preserved ration which is most commonly found in tray rations and is heated by immersion (nicknamed T-Rat).

- Meal, Ready-to-Eat; The standard individual field ration (commonly called the MRE).
- First Strike Ration; An individual ration designed for being edible while on the move during assaults.
- HOOAH! Bar; An energy bar which is the successor to the D-Ration and is found in some MRE menus.

As previously stated with the MREs, it was '*reportedly*' a self-contained individual field ration. Reportedly is an important key word. MREs are *dehydrated* meals, and therefore, require water in preparation *before consuming* them. Therefore, MREs are not fully self-contained as advertised. The requirement of water for consuming the ingredients *is a serious downside* issue of the MRE. Imagine being lost at sea or in the desert with very little fresh water. MRE contents can be ingested without water, but will greatly increase the need for intake of water. Whereas, C-Rations (or MCIs) contain water in their ingredients and are sealed in cans. The other downside issue with MREs is the packaging. It can more easily be ripped open as compared to the tin cans of the C-Rations (MCIs).

Again, as Napoleon Bonaparte once said, 'An army marches on its stomach' – so must an army's kitchen keep pace with its marching. As the world's weapons technologies become more and more complex, so must the warrior in the conflicts. Thus, military food service quartermasters and suppliers must keep pace with advancing culinary technology, as well as packaging techniques.

CHAPTER XV
DUAL-USE MILITARY TECHNOLOGY

"Ya know guys, this here truck would make one great SUV when I retire!"

Military technology which has application in the private sector, or civilian world, is called military dual-use technology. The greater segments of those developed technologies are *usually* maintained within the military system for a number of years before being released into the civilian world.

Around 800 BC, the Assyrians introduced solid foundational concepts of logistics for supplying their imperial troops. Some of those efforts became a significant component of modern business for the allocation and provision of supplies to meet needs – in the field of logistics. Around 600 BC, the concept of a postal service was introduced as Persian emperors sought to maintain communication with field commanders. Around 100 BC, Chinese armies began using the wheelbarrow; a new piece of technology so vital for transporting military equipment its design was kept a military secret for many years – by the emperor's mandate.

For conquering their known world, the Romans developed a massive system of roads. Thus, from that huge arrangement of interconnecting highways the cliché of 'all roads lead to Rome' was born. The Roman road structure ultimately provided basic blueprints for the modern superhighway system of today. Its concept was introduced in the 1950s by President Dwight Eisenhower – and, interestingly, primarily with military transport needs in mind.

Military ingenuity goes beyond the study of various applications for use in increasing military might. Its creativity may also extend into methods for producing military equipment and ways for improving performance, while also reducing material and technical requirements for all phases of production processes. During WWII, Germany expended a tremendous effort for producing artificial rubber and fuel to reduce, or eliminate, their dependence on imported petroleum, oil, lubricants, and rubber supplies. In accomplishing this massive effort, Germany's military scientists maintained meticulous records and manuals. Some of those records and manuals were used for making great technological improvements in society after the war ended. Just one of those technologies they were pioneering was headed by Dr. Warner Von Braun, and ended up becoming what is known today as the U.S. National Aeronautics and Space Administration (NASA).

Therefore, military records and supporting tech manuals are very important for having dual-use technology!

Military ingenuity may be unique only in its *specific military application*, and NOT in its use within its basic scientific and technological achievements. Because of this uniqueness of specific use, military studies strive to incorporate self-generating evolution levels within their design applications. Therefore, military manuals can be updated regularly with new and improved technology information and procedures.

In general, military manuals are great technology resources, resources which troops appreciate (and even people not in the military should appreciate). The most common complaints about military manuals are that they are hard to read, boring, and poorly written. Because they are generally written for field use, they are not scripted for being prized pieces of literary prose. Military manuals are written for logically getting points across and getting jobs accomplished with the least negative impact. Since they are not great works of literature, they were never intended to be such. Military manuals are actually working notes which ended up as guiding instructions for specific procedures.

Military manuals are loaded with vast amounts of technology information, accumulated with great difficultly, and often at the cost of many lives. Great amounts of effort, vast amounts of experience, time, thought, and usually funding, went into producing military manuals. Military manuals generally have short life spans, with field manuals usually being replaced every ten years. Technical manuals are in print as long as the applicable piece of equipment is in the military inventory. Numerous individuals find military manuals very useful. This of course not only includes military personnel, but also historians, law enforcement personnel, collectors, museum curators, scientists, survivalists, and war-gamers.

For all military personnel, military manuals are a repository of information which can often prove to be useful in many circumstances. Manuals are their guidelines, which coordinate them with other soldiers and old manuals, and can supply them with many useful miscellaneous techniques which are not commonly used.

For the historian, military manuals are an invaluable statement of doctrines of the time for when they were written. They provide a record for the capabilities of equipment and how everything was supposed to operate.

For law enforcement personnel, military manuals are sources of useful combative and defensive techniques which can be selectively applied.

For the collector and museum curator, those old manuals are often the *only record* remaining on how their relics had operated, were to be maintained, and were to be repaired.

For war-gamers, military manuals provide the over-all guiding details for how equipment is suppose to function.

For scientists, military manuals are found to be useful as originating points in research for documenting what is already known about a process, or was once known.

For researchers and scholars, they are vast pools of knowledge which can be used to fill in the many pieces of complex puzzles possibly missing in their subject area of expertise and study.

While many technologies invented for military use involve weapons, many other military inventions were developed by doctors, design specialists, and culinary personnel in support of military operations. The act of inflicting harm has actually propelled medical science into developing new techniques for treating injuries, while others have excelled their resources in other areas. Over time, these non-destructive inventions found their way into the private sector, where the civilian world has enjoyed the benefits of newly discovered technologies.

Follows, with some detail, are three military dual-use technologies which has significantly changed the world we live in:

Food Preservation – The practice of preserving food was first used by General Napoleon Bonaparte as the French military forces needed large amounts of food for a growing army. It was the French who, at first, began using champagne bottles for preserving their food. Then later, the English troops began carrying metal cans of food as they engaged in military campaigns around the world. The English public then started buying canned food in 1830; preserved food was at that time, however, expensive and considered a luxury for the wealthy.

Plastic Surgery – WWI involved large amounts of troops fighting in trench warfare and using more destructive weaponry such as bombs, grenades, and machine guns. Sir Harold Delf Gillies was widely considered as the Father of Plastic Surgery. Sir Gillies began pioneering facial reconstructive surgery because the Doughboys were often hit in the face after popping their heads up and out of the trenches in an effort to see what was going on with the enemy forces. American doctors Varaztad Kazanjian and Vilray Blair were employed by the U.S. Army during WWI specifically for plastic surgery. They leaned from Sir Gillies' surgery achievements and expanded his work. Both artists and doctors worked together trying to restore the facial features of injured GIs and veterans of that war and on into WWII. The Smithsonian website documents many of these achievements.

Trench Coats – The fashion world owes a great debt to the military for the invention of the ageless trench coat. Thomas Burberry invented a water-repellent material of which trench coats were originally made, and which the British military issued to its officers. The trench coat came by its name literally since it was used by British and U.S. soldiers in the trenches during WWI. The shoulder loops on the trench coats held the rank insignias of the wearers. Trench coats were lightweight, allowing the wearer to stay warm while moving more freely as compared to heavier coats that soldiers had been previously issued. Later in the 20th century, civilians began wearing trench coats as a fashionable trend.

Military dual-use technology developments rapidly multiplied right up to the U.S. Space Program, as well as those of other countries in the late 20th century – and it is continuing to proliferate on into the 21st century. The following is a *very brief* list of dual-use technologies which the military invented, developed, and was transitioned into the civilian world. Its benefits have changed the civilian world forever.

Again, this list is very short; the complete list is actually *quite long and grows longer every day…*

- Catapults
- Gun powder
- Cannons
- Trenchcoats
- Logistics
- Wheelbarrow
- Postal Service
- Superhighways
- Weapons and munitions
- Jet engines
- Radar

- Sonar
- The ID card
- Quick clotting powder
- Better clocks and time pieces
- Global Positioning System (GPS)
- Atomic power (and the atomic bomb)
- Solid and liquid fuel rockets
- The space program
- Polypropylene (thermal) underwear
- Night vision devices
- Earth orbiting satellites
- Walkie-Talkies (field phones)
- Digital photography
- M&M Candy
- The Jeep
- Computers (a decoding device)
- Radiophones (forerunner of mobile phones)
- The Slinky Toy (a new type of spring)
- The Silly Putty Toy (intended to be C-4)
- Microwaves (and microwave ovens)
- The laser
- Global weather forecasting
- Robots
- Velcro fabric
- The Light Gun (used with video games)
- The INTERNET (a derivative of MILTNET)
- The '@' as a symbol for internet address use
- Erosion resistant coatings
- Hazardous material encapsulation
- Rapid adhesive bonding
- Bio-based hydraulic fluids
- Permanent marking techniques

CHAPTER XVI
TECH MANUALS

"Guys, this cannot be serious!
We forgot to pack the tech manuals???"

The first use of military manuals will never be known, but historical records trace their usage back for many centuries. Historical records in art form exist which date back to the 5th century BC crediting an Indian named Han Lo-Ming for creating Chi Hsuan Men (an 'Unusual Style' of martial arts). This artwork used defensive scissor techniques of the White Jade Fan for trapping swords and spears, and pressure point strikes with the fan's tip. Chinese historians dispute India's claim to being the originator of Asian Martial Arts. They point to military manuals and documents dated from 206 BC to 220 AD. Their claims are that Han emperors actively funded the study and refinements of Kung-Fu far beyond any fighting system known in India during this period of history.

Some earlier examples of historical martial arts consist in a series of images only. The earliest example is the wall mural in Tomb 15 at Beni Hasan. It shows illustrations of wrestling techniques which date to around 2000 BC. Similar depictions of wrestling techniques are also found on Attic vases dating to Classical Greece.

The only known instance of an actual 'technical manual' from Western antiquity is from the 2nd century which details Greek wrestling techniques. There are some examples in classical Chinese literature which predate the turn of *Anno Domini* (the AD years), one of which is *The Records of the Grand Historian* by Sima Qian (around 100 BC). It documents wrestling and refers to the earlier how-to Han manuals of the (2nd century BC). Another Chinese text on wrestling is *Six Chapters of Hand Fighting* and is included in the *Book of Han*.

All other existing military-like manuals date to the Middle Ages or later. The *Combat Stele* at the Shaolin Monastery dates to 728 AD. The earliest of those texts detailing Indian martial arts is the *Agni Purana* (about 8th century), which contains several chapters giving descriptions and instructions on fighting techniques. The text describes how to improve a warrior's individual expertise for killing enemies using various methods in warfare; whether they went to war with chariots, horses, elephants, or on foot. The oldest existing European martial arts manual is the *Royal Armouries MS I.33* and is dated around 1300 AD.

The exodus event in the Book of Exodus in the Holy Bible is considered by most scholars to have occurred around 1446 BC, with its authorship attributed to Moses. As to when Moses finally penned the very first Book of Exodus, that original date will never be certain. However, the Book of Exodus does contain some basic military field guidance. In addition, other books of the Bible contain some basic military field strategies. Therefore, it could be stated that the Bible contains some limited military information as a military handbook-like manual. Since the best age estimate for the Book of Exodus is 1446 BC, then it could be stated the Bible is another example of a limited-use military-like manual with sections of it dating as far back as 1446 BC.

As previously stated, the first use of military manuals will never be known, but historical records trace their usage back for many centuries. The Ancient Roman Empire was perhaps most likely the *best example* for use of military strategy and manuals. Including the old Roman Republic, the Roman Empire existed for 984 years, or either 1961 years (508 BC to 476 AD, or to 1453 AD), depending on how its final end is viewed. The Western Roman Empire fell and ceased to exist in 476 AD, but the Eastern Roman Empire continued on until it finally fell in 1453 AD.

Regardless of how its length of reign is viewed, it was nearly one to two millennia. Such a long length of reign in an enormous might of power required an excellent military system of logistical processes as much as possible (manuals included). One event, however, propelled them into engraining strategy and manuals within their military machine more than ever – it was called the Battle of Allia (so named for the Allia River near Rome).

In 390/387 BC, Rome received its greatest humiliation, as the Gauls (primarily the French) ransacked the city. This battle (and defeat in the Roman Capital) marked the need for more detailed documentation of strategy and logistics. If Rome was to reestablish its authority as the central power of Italia (Italy), and be prepared for meeting possible similar future disasters, then some needed reorganization was definitely required. The use of military manuals became one of the more significant changes in military strategy operations. These changes were accomplished in greater effort by succeeding Romans, and a certain Hero of Rome named Fluvius Camillus was given much of its credit.

The Eastern Roman Empire (also called the Byzantine Empire from 330 to 1453 AD), for much of its history, was one of the major powers of the medieval world. As the Western Roman Empire declined and fell in 476 AD, the Byzantine Empire became the inheritor of its centuries of traditions and institutional policies. Throughout most of its entire history, the Byzantine Empire was assailed on all sides by various superior enemies. Therefore, the empire maintained a highly sophisticated military logistical system, which relied on discipline, training, knowledge of tactics, and a well-organized support system.

Crucial elements in the maintenance and spreading of this military knowledge, along with tradition and policy, were kept in their various military manuals. Byzantine manuals began greatly proliferating into the 10th century when the Byzantines embarked on their conquests in the Eastern world and into the Balkans. Then, early in the 11th century, their manual productions began decreasing. From the large amounts of Byzantine military documents which survived the centuries, it is evident that their soldiers relied heavily on their manuals.

The value that military manuals could be for historians was previously stated in an earlier entry. Thus, knowing how methodical and meticulous the Romans were with their military manuals, this casts a qualitative dimension to an interesting piece of military correspondence that was discovered from the Western Roman Empire. The correspondence was military documents regarding a certain religious zealot from about 2000 years ago. Those certain documents were official government correspondence between the Procurator of Judea (Governor) and the Emperor of the Roman Empire (Tiberius Caesar). The correspondence was in reference to Governor Pontius Pilate reporting the crucifixion of one Jew named Jesus Christ, troubles with His execution, the fact that He rose from His tomb, and was reported seen alive many times again after His death by officially witnessed execution.

The letters of correspondence were actually Government Reports from Pontius Pilate to Tiberius Caesar, and were regarded as genuine by the early Church Fathers. The letters go into great detail about the circumstances of the crucifixion of Christ and supernatural events immediately following Christ's death. In these reports, Pontius Pilate (an eyewitness to some of these events), relates many of the miracles performed by Christ. Most importantly, thought, were descriptions in detail of the paranormal occurrences at the time Christ was crucified.

Many references to this military correspondence can found in numerous books, one of which is titled, *The Lost Books of the Bible,* and was published by Testament Books of New York. German Bible scholar Dr. Constantin Von Tischendorf (1815-1874), possessed in his own personal library, copies of this correspondence from antiquity which had been copied in Greek. His personal copy had also matched the Syriac manuscript copy. Thus, the second antiquated manuscript (the Syriac copy) further added creditability for verifying older known manuscript's authenticity.

These oldest known copy of this military correspondence can today be found in the British Museum in London. Interesting how important some military manuals (and/or military documents) ***can be*** for proving and validating certain 'questionable events' – but ***are not even considered by the masses***.

So very *logically, truthfully,* and, *profoundly,* interesting – isn't it?

CHAPTER XVII
GOD'S TECH MANUAL

All during my own times in the military, everything I did was accomplished through some manual, with supporting directives. In the maintenance community, those manuals and directives were published and issued by what was generally called the Cognizant Field Activities (CFAs), Depot System Commands (DESCOMs), Air Logistics Commands (ALCs), etc. CFAs obtained their technical data directly from the Original Equipment Manufacturer (OEM). The OEMs were staffed with engineers of various skills who designed and built the military equipment, and then tested their limits to sustain the operational demands needed by the various field activities.

When performing maintenance on military aircraft, I was to perform those maintenance tasks exclusively by the Aircraft Maintenance Manuals specified by the CFA or for that particular Type, Model, and Series (TMS) of aircraft. Any deviation from those specific 'tech manuals' would most likely result in an improperly maintained aircraft. That, then, would result in a very poor quality weapon system, thereby degrading fleet readiness and thereby degrading national security. Improper maintenance would also most likely result in an aircraft mishap, resulting in millions of dollars of weapon system loss, and unknown numbers to loss of life. Therefore, there are many reasons for following tried-and-tested maintenance and repair procedures by the book. No other tech manual would be applicable for the unique systems in the aircraft TMS, which I was maintaining or flying in.

Such is similar with the Bible. Someone once said and wrote that the letters B-I-B-L-E could stand for 'basic instruction before leaving earth.' That does sound excellent, especially to those who come from a 'tech manual' background. God inspired several men, through His Holy Spirit, to write the Bible, the Book. God is the Creator of the universe. He made it all; He knows how it all works. He made mankind, and therefore, God, as the OEM of everything, knows more about the original design of mankind, inside and out, than any of the rest of us. God's detailed, intimate, and thorough knowledge of humanity is past, present, and future.

Doesn't it, then, make logical sense that the Bible be consulted for guidance regarding the issues of life? Why not go directly to the writing of humanity's OEM for the answers to all the concerns of life? The Bible contains lessons on national as well as individual levels, yet applicable for any level of society and culture. The Bible contains answers to easy issues and hard ones as well. The Bible is full of history about peoples making some serious mistakes and rejecting God's help. The Bible is full of history about peoples making some serious mistakes and accepting God's help. The Bible shows the results when we 'do it our way,' and it shows the results when we do it with His help, His way. The Bible is a tech manual, an operator's manual, and a guidebook, which God gave to the people He made—us. The Bible is also God's love letter to His prized creation—us.

Precious metals and stones have to be mined, and that mining comes with patience and diligence. So do answers from the Word of God; they must be mined with diligence and patience, for it is more than just any book. The Holy Bible is a spiritual guidebook: *"Your word is a lamp to my feet and a light to my path"* (Psalm 119:105). As the answers come and are revealed to the patiently diligent seeker, those answers sometimes become so profound their revelation becomes imprinted within the heart and mind. Those imprints are personal imprints for that one who sought truth from the Lord God. No one can then steal those imprints away from the heart of seeking pilgrims in their journey of life.

With those precious imprints, that one will always have a map for traveling the road of life, a technical manual for navigating its difficulties, and even a construction manual for expediting the repairs with the walls we drive through scattering shrapnel everywhere!

CHAPTER XVIII
THE ILS PROCESS

"Ah-oh, somebody failed on their ILS planning and we're in trouble big this time..."

Confucius once said, 'To put the world right in order, we must first put the nation in order; to put the nation in order, we must first put the family in order; to put the family in order, we must first cultivate our personal life; we must first set our hearts right.' Likewise, to manage life's gauntlets, ups, downs, failures, successes, and so forth, one must follow a similar basic concept. This concept is especially effective with putting a disarrayed life back in order.

James Yorke once said, 'The most successful people are those who are good at plan B.' With those words, the next topic comes into view. When accomplishing anything successfully, there must first be a plan. That plan includes projecting preliminary preparations, developing action policies, and devising contingency strategies. The preliminary planning efforts examine the forecasting of future issues. The action policies examine forecasting the action steps needed for accomplishing the desired goal. The contingency strategy efforts examine the forecasting of results from failed actions, potential unexpected events, and then put in place a guide for dealing with those issues. Every successful plan for accomplishing anything must contain ***the minimum of those three aspects:*** (1) Preliminary Preparation Forecasting (2) Action Policies (3) Contingency Strategies.

The United States military establishes plans utilizing this format. This planning format is called *Integrated Logistic Support*, and its termed acronym is ILS. ILS is the disciplined and unified management of all the technical and administrative disciplines that plan and develop logistics support requirements for military forces, and which will ensure system product quality in terms of reliability, availability, maintainability, and testability. The ILS process is a major framework factor in all planning for any type of fielded weapon system along with its support program. Embedded within the ILS Process, are the logistics disciplines of: (1) Reliability Engineering (2) Supply Support

(3) Test Equipment (4) Manpower & Personnel (5) Training (6) Technical Data (7) Computer Resources (8) Facilities (9) Packaging and Transportation (10) Design Interface

These ten planning elements make up the field of Integrated Logistic Support, or ILS, which is an extremely effective project management program tool.

The founder of Federal Express, Fred Smith, once stated, 'You are the way you are because that's the way you want to be. If you really wanted to be any different, you would be in the process of changing right now.' There is no lightning-fast plan or program for repairing defects in human life. We are complex and, therefore, some of the repair patches are complex as well. Today's society seems to desire operating more at a rate of hyper-baud rather than super-baud when connecting with anyone. And, that even includes the Lord God.

Speedy relationships most times do not equate to lasting ones. Relationships require time with their connections for maintenance and sustainability. This very same principle applies to a relationship with God. Personal time must be invested with God for a sustaining relationship – end of subject.

Astronomer Carl Sagan was probably the most instrumental person in introducing our world to the expanded universe. He commented that there were billions and billions of stars out there. So, then, consider this: of all the billions of solar systems in all the millions of galaxies in this universe, why would God want to fix a broken down cosmic sphere such as this one? Yet He did. This backwater planet we call Earth is becoming wasted by its inhabitants. Yet God still desires to be in contact with His prize creation He entrusted it to – humanity. God chose in sending His Prince, His only Begotten Son, to give His life as ransom, so that all the corrupted inhabitants of this planet could go free – but only if they chose to accept His sacrifice.

Interestingly, Carl Sagan is quoted as saying, 'Science is not only compatible with spirituality; it is a profound source of spirituality.' With some intellectuals now of the opinion that science and spiritually have a connection, *then* God surely can change a life who desires it. Mahatma Ghandi once said, 'Be the change you want to see in the world.' By applying those wise words to a relationship with God, change is inevitable. The contents of this book address the basic issues of life. Absorb its 'ILS' suggestions, and let them be the catalyst for change. Then with God, let His Spirit dramatically change whatever is needed in your life.

Chuck Colson once stated that the most quoted verse in the Bible is, 'God helps them that help themselves.' This 'verse' has been falsely attributed to being in the Bible, and its words have also been falsely attributed to Benjamin Franklin. Mr. Colson went on to say he found out God helps those who ***cannot*** help themselves. Those who cannot help themselves at a minimum have the simple spiritual road map of the Ten Commandments found in the Bible for guiding them along the way of life.

Interestingly, the basic issues of life that everyone deals with, requires a discipline structure similar to that of the military's ILS system; and will effectively manage life's intriguing facets. The basic life disciplines are: laughter, friendship, faith, prayer, forgiveness, loneliness, words, love, trust, attitudes, perception, choices, and character. Also interesting is that those basic issues of life disciplines are collectively dealt with, and governed by, just ten simple principles. Those ten principles are the Ten Commandments (mentioned above) as found in the Holy Bible.

A minister once said that if everyone would live by the Ten Commandments of the Holy Bible, there would be no crimes. He went on to say all the issues of life are wrapped up within the fabric of those ten social policies. Those Ten Commandments and their associated precepts are the starting points for truth in exploring other facets within life, and also this book. The Ten Commandments are the floodlights for the road of life we all travel. They become the lighthouse of guidance broadcasting a beacon of wisdom into every heart that will absorb its virtues. The Ten Commandments from Chapter Twenty of Exodus are as follows:

(1) You shall have no other gods before Me; (2) You shall not make a carved image and bow down to [worship] them; (3) You shall not take the name of the Lord your God in vain; (4) Remember the Sabbath Day, to keep it holy; (5) Honor your father and mother; (6) You shall not kill [murder]; (7) You shall not commit adultery; (8) You shall not steal; (9) You shall not bear false witness against your neighbor; (10) You shall not covet your neighbor's house [nor anything he has].

These Ten Commandments are the basic traveling laws for governing a safe passage through this journey called life.

When traveling anywhere, directional signs are encountered. If by land, there are road maps with map indicators and actual road signs. If traveling by sea, the mariner has sea maps with sea currents, shipping lanes, latitude/longitude, marker points, and celestial data. If traveling by air, the aviator has aerial flight maps with flight levels, beacon points, marker points, vectors, and celestial data. When traveling anywhere, in getting there we have prepared visual aids for assisting our arrival. We must, though, strictly adhere to the directions provided. If not, wrong turns be made and passages can be taken resulting in becoming lost. More seriously, accidents can be caused resulting in loss of life and property.

When traveling anywhere, prepared visual aids can greatly assist in our arrival. However, we must adhere to those directions provided, or we may arrive much later than intended at our projected destinations – or possibly never arrive at all. When directional signs express stopping, yielding, no passing, etc., those signs mean exactly what they express. Some Traffic Engineer somewhere studied quantifiable data for design and placement of those directional signs. Some Traffic Engineer examined the terrain, the prevailing conditions, the regional laws, and concluded the course of actions needed for the correct data to be imprinted on those signs.

Those directional signs were installed for safety and guidance in directing travelers to their destinations. Any deviation from them will result in not arriving at the desired locations. The traveler could end up lost, confused, and possibly sometimes in unsafe situations and/or locations. Strict adherence to directional signs while traveling is a ***must***.

Many years ago when I first became a Christian, the Lord God impressed on me once while in meditation, "If you follow My word you will never be lost. Walk with Me through your life." Those words of His impressed within me were never forgotten. All through life since the time of 'hearing' them, I have tried to follow God and walk through this life with Christ. It was through those daily walks of faith with God and His Son Jesus Christ that I discovered being found from being lost. It was during those travels with God that His Spirit became the road signs guiding me through life's traffic congestions – and away from more accidents than I could have had.

Many times down through the years of walking with Christ, lyrics of the song *I'm Using My Bible for a Roadmap* have echoed within me. Many outstanding singers have performed that popular bluegrass gospel song. Its words relate to the Bible being the Christian's roadmap with no detours on the heavenly highway. The lyrics also state the Ten Commandments and the Twelve Disciples are guides along the way for the trip. Psalm 119:105 says, *"Your word is a lamp to my feet and a light to my path."* Thus, the Bible tells us within the text of God's Word that great illumination exists for life's travels.

In Psalm 37:23 is found, *"The steps of a good man are ordered by the LORD..."* Thus, the Bible tells us if we live in fellowship with God, He will guide our travels of life. This passage of Bible Scripture is commonly quoted, but uncommonly quoted is the very next verse of Psalm 37:24, *"Though he fall, he shall not be utterly cast down; for the LORD upholds him with His hand."* Thus, the Bible tells us if we live in fellowship with God, even though we falter and fail, God will not reject us like everyone else can, and usually does. He will still pick us up, and He will faithfully continue guiding us along our travels of life.

In doing so, even in the midst of complete meltdown failures, God's Holy Spirit comes near. God sends His Holy Spirit to His broken people as directional signs for returning to Heaven's Highway. Sometimes those signs are marked with such displays of assistance that the receiver knows only God could have provided. Those were 'signature' signs personal between the receiver and the Lord God. Most times, though, those signs are marked with the presence of God's Spirit, and are accompanied by tears. Those tears announced the workings of His Spirit, for helping us wash away the inner stains of the wrong turns we have made. Those signature signs stamped with tears are God's imprints within the hearts of each receiver. Those imprints stay tucked away for exposure to others as illuminating directional signs along life's inroads toward Heaven's Highway.

In Chapter Six of the Book of John, Jesus was teaching of spiritual communion with God the Father; that no one could come to Him without God's Spirit drawing them. From that time, many of Christ's followers turned away, and He asked the remaining Twelve Disciples if they, too, wanted to leave Him. In verse 68, the Apostle Peter replies, *"Lord, to whom shall we go? You have the words of eternal life."*

God's words and impressions can become directional indicators within the heart and mind of each person to whom He sends them. They know from Whom that guidance comes, even though the rest of the world may laugh, and scorn, and turn them away. God's personally impressed directions will line up with the Bible, His Word. The Bible will lead people in the right direction; and if they follow the correct road signs, it will lead them home to God.

Light is often used as a metaphor for truth, understanding, and guidance. The Light of God can shine into the deepest hidden crevices of the human heart and soul. That Light can reveal buried and (sometimes forgotten) broken pieces in need of repair for securing a brighter future on the horizons of tomorrow. That Light can also illuminate what appears as unintelligible, into quantifiable understanding. His Light opens the doors of new dimensions for comprehension, and that of the spirit world as well.

Just as the military's ILS Program can successfully guide, build, launch, and field a new weapon system – so does the Holy Bible with those willing to be led by it. The Bible will guide, build, and field a new Christian life in this on this side of eternity. God's Light provides an Integrated Life Support program for the Issues of Life Study-map needed along this journey of life – His ILS is required for our ILS. Once we have connected, He will then successfully spawn its launching into the eternity of the next life with Him.

The simplest of all life's directional signs for finding Heaven someday is just turn Right, and go Straight. God's Light will always brighten the passages we traverse.

"For with You is the fountain of life; in Your light we see light" (Psalm 36:9). *"Oh, send out Your light and Your truth! Let them lead me; let them bring me to Your holy hill and to Your tabernacle"* (Psalm 43:3). *"The entrance of Your words gives light; it gives understanding..."* (Psalm 119:130).

A Christian Codex (book) from the first century with metallic pages. Evidently, Christians of antiquity felt the Gospel message was so significant that, when possible, they would inscribe its message on metal plates and assemble them into books. This practice made 'books' more durable than scrolls, and illustrates seeing the need for long range planning with carrying the Gospel of Christ's message far into the future.

CHAPTER XIX
ENDING CREDITS

"Think we're done here, but somebody else's gonna get the credit for this one"

Have you ever done something you were never given proper credit for? Yes, we have all fallen victim to those situations. What governs how we live out those events, though, is the way we deal with the 'end credits' of those scenarios.

End credits are most commonly known as those listings of recognition passing in review at the end of theatrical productions or movies. Many viewers do not pause for viewing the end credits, and thus, the supporting staff and crew responsible for developing, assembling, and producing the movie go by largely unrecognized. Sadly, too many times it is only the main actors in the cast who are recognized and remembered for their work in the movie. Such is similar in many areas of life. Far too many times, people fail in properly recognizing the end credits in the lives of other people's endeavors.

There was once a time while I was assigned to the FRAG mentioned in the previous entry, that a serious incident of 'end credit piracy' occurred. Select details and timelines of the incident will be sanitized so as not bringing embarrassment to certain agencies.

For several months one year, a certain aircraft oil pressure indicator became in short supply in one military supply system. This shortage resulted in the downing, or grounding, of many aircraft of certain TMS around the world. This rippled into the effect of limited mission capable aircraft, and then for mission flight sorties. The end result became partial loss of mission readiness for one facet of that military service.

All of the fleet squadrons I supported were contacting me on a daily basis pleading for availability of those oil pressure indicators. After months of this scenario and mission readiness failing to a critical level, I began 'thinking outside the box' as the saying goes. Across the hall from the FRAG Office was one of base engineering technical libraries, and into its files I went.

Within those files, I dug up the technical drawings for the oil pressure indicator the fleet so badly needed. Oddly, in those drawings was found two other oil pressure indicators with the same part number, but different dash numbers. Further research revealed one of those other dash numbered indicators was identical to the one my service needed, with one exception – the external canister was one-half inch longer.

Having been a former Aircraft Electrician, worked on that TMS aircraft many months, and removed and replaced that particular indicator many times, I was very well acquainted with its location on the instrument panel in the aircraft cockpit. It was my preliminary assessment the other dash numbered oil pressure indicator could be a viable substitute for the one my service desperately needed.

Going to the FRAG supply computer terminal, I found the national stock number (NSN) for the particular dash part number of the oil pressure indicator I'd discovered on the technical drawings. Then, bypassing my military service's supply system, I began researching the other military service's supply systems. Another service had over 400 assets (indicators) in 'A Condition' (ready for installation) in their supply depots. Their low reserve limit was 200, and thus, if authority was granted by the Item Manager and Cognizant Engineer of that indicator for both military services, my service could 'borrow' up to 200 of them.

At the time, my service was in need of about 125 'A Condition' assets. Therefore, our critical fleet mission needs would not degrade the criticality of the supply for mission needs of the other service. It would be a win-win scenario in my technically non-engineering opinion. To my supervisor my findings and recommendations were immediately brought. After hearing my explanation, he immediately rushed me into our duty truck and drove me to the base engineering office.

The next afternoon, I received a phone call from the technical branch of the supply office of my military service. The representative informed me their office and our base's engineering office had reviewed my findings and recommendations. Everyone fully concurred and subsequently the other military service's supply channels had been contact for support and approval of 'borrowed' assets. Support and approval had been granted and a Technical Bulletin was being drafted that afternoon allowing substitution of the alternate part numbered oil pressure indicator I'd discovered.

The representative asked me to wait on the line for a moment because someone else wanted to speak with me. The next individual was one of the chief engineers for the project. Both representatives expressed their sincere appreciation for my diligent effort with attention to meticulous detail, without which this project would not have been possible. Additionally, my efforts directly resulted in a rapid change of mission critical flight status for many squadrons around the world for my military service. Within a matter of only about two days, literally several dozens of mission critical flights were undertaken again, and things were well on their way to becoming normal – well, that is for the U.S. military setting things straight in harm's way.

About a week later, in response to all this, I received a very nice letter of appreciation from my commanding officer. Subsequently, weeks later some of the squadrons I supported sent me letters of appreciation signed by their commanding officers. However, about a month later, rumors began flying around about a significant Beneficial Suggestion (BE) being awarded at the supply office of my military service. The BE was for an alternate oil pressure indicator, and was being jointly awarded to the two men who talked with me on the phone about a month prior.

Immediately, my supervisor and I placed phone calls to both of those individuals, and no responses were returned. Contact calls were made for several days and no response calls to my supervisor and I never were returned. There was a very informative media article in the next quarterly military magazine generated by the command of which that military supply unit was attached. The article contained information regarding how those two individuals who had talked with me on the phone, had researched and resolved the supply and logistics problem for the much needed oil pressure indicator for their service. The media article went on mentioning many had accolades from their commands for unselfishly contributing to their country and the support of the military services during a critical time of needed defense. One of the most eye popping catchers was the article line mentioning the BE the individuals were awarded and shared along with a photo. The photo was of them and a large representative check. The check was in the monetary four digit range (it was indeed very significant, but I'll not mention the amount).

For me, though, I was in the throes of resolving the problem 'in the trenches' and placing backorder requisitions for my squadrons to be first in priority for shipment of those 'A condition' assets when they became approved and available. For me, I had downed aircraft all over the world needing to be placed in mission-critical status – and that was my number one PRIORIOTY. For me, I was not afforded the luxury of time to submit any paper documentation for my discovery.

Saving the day for national security and doing my job cost me significant notoriety, not to mention several thousand dollars. However, saving the day for national security and doing my job won me the favor of many in the field. It may never be known for sure the full extent the blanket of protection those mission ready aircraft helped to provide for freedom in the far corners of the earth.

Such are the REAL end credits; knowing one did their job and made a REAL difference in the world. Real end credits make a better and safer world; they don't just line the pockets of its characters.

CHAPTER XX
SWITCHOLOGY

"Think somebody said there may have been some errors with of the several switches?"

There is an illustrious term all of the military maintenance technicians love using. It is just simply 'switchology' and it is short for getting all the switches in the right order! If the operator of any high-tech weapon system does not have certain switches selected for appropriate modes of operation, systems will become 'inoperative (inop)' to 'not working as advertised' according to the technical manuals. Switchology can result in what is sometimes called 'selective malfunctions.'

This is why reading and being thoroughly familiar with the Operator's Manual for each weapon system –AND– receiving written and practical examinations are *absolutely* essential for operational safety and mission readiness.

Many times the younger-aged, or inexperienced, or even the down-right overconfident (cocky) operators of aircraft, tanks, submarines, et cetera, have misjudged what specific switches were supposed to have been placed where, at what time, and for what select purpose. Then, something goes inop, or maybe somewhat awry, to even totally blamo at the wrong or inopportune times. Then there will be much explaining to do – IF the operators are alive for returning to home bases and reporting directly in front of their unit commanders.

Many times as an Aircraft Mechanic, Electrician, and an Examiner, when receiving write-ups (gripes, squawks, red-lines, gray-lines, et cetera) from pilots and crews, I'd first check the Operator's Manual for the specific system, determine the mode of operation the operator was attempting, and replay the switch order placement as prescribed by the technical manuals. Many times discrepancies could be discovered in error of switch misplacement. I'd then sign-off the maintenance forms (as did many technicians) with 'Switchology Error; Next Pilot/Crew Check' which meant I found the probable cause of error was in switch placement and for the next pilot and/or crew to be sure and recheck that system's operation 'by the book' (the Operator's Manual).

As much as I've signed off with 'Switchology' for maintenance troubleshooting, I have even been guilty of doing things similar to that myself! Hey, we're only human as has been said. We all make mistakes.

Once I was running up a twin engine T-2 Aircraft (jet) on the high-turn area of an aircraft tarmac. At the time, I was a Petty Officer Second Class (E-5) and was in the pilot's seat (front cockpit). My Third Class assistant (Jayse) was in the back seat. That evening we had previously completed removing and replacing an air-conditioning turbine in the aircraft. When accomplishing the maintenance task, sometimes gasket sealant at the line clamps could seep into the duct lines at the clamp joints when the clamps were tightened. When a little too much sealant had entered the ducts, an odd odor would appear in the cockpit air when the air-conditioning system was turned on (hot or cold).

The quickest way to remove this odor from the system is to run both jet engines at full throttle for about thirty minutes or so while the air-conditioning (A/C) system is selected to 'full heat.' This 'burns' off the sealant from the duct lines at the clamp joints, and saves a tremendous amount of maintenance man-hours. However, the aircraft operator has the responsibility for 'knowing' the aircraft system's operability while in the pilot's (or operator's) seat.

The aircraft only had about half a tank of jet fuel when we began our run. Then after checking the

A/C system, the odd odor was discovered, and a little stronger than what we'd like to have smelt for having the half tank of fuel. By the time the odor had vanished about forty-five minutes into the 'burn run,' the aircraft was standing quite tall on its landing gear struts. After shutting the engines down, and accomplishing the required cockpit post-flight check, I engaged the canopy switch to 'Open.'

It did not move at all! Its actuator did not even whine!

Looking back at Jayse, he looked at me and we both exclaimed, 'Main Landing Gear Safety Switch! How could we forget it!'

That switch was a safety switch for not permitting the canopy to be inadvertently actuated open while the aircraft was being taxied around the tarmac and the runway prior to takeoff. The low fuel load had brought the aircraft profile higher up on its landing gear struts to where the safety switches had become actuated. There both of us sat in the aircraft cockpit; the ground crew already driven off, and we were stuck inside the plane. Woe be unto us! With the aircraft sitting tall on its struts, that ridiculous situation brought an entirely new perspective to the cliché of 'Jacked Up'!

After a few minutes from the backseat came, 'Adams, what now? Our buds are gone and we're locked in out here.'

'Yeah, I know. Gimme a minute and I'll think of somethin'' was my response as I looked through the canopy at the darkening night skies befalling us.

From the back seat came, 'Hey, man, like this plane's engines are switched off and heat's buildin' up. It could catch on fire and we'd be inside. That ain't so good.'

'Would you just shut up back there' I said back to Jayse. 'By the way, the engines of these planes are switched on and off several times a day and they don't go boom.'

'Yeah, but ya know, it'd be our luck for us bein' the first casualties of a flight line high-turn melt-down now wouldn't it?' came back Jayse.

'Ya know Jayse, them negative vibes all the time ain't doin me any good up here! Just shut up so I can think a bit' was my reply.

After actually getting a clear minute for thinking, it hit me what to do. No worries! I turned the Battery Switch back on and turning on the aircraft radio, squawked the squadron Line Shack. Embarrassed, I asked for the Line Boss. He came on and said, 'Adams, don't tell me you need some fuel. Ya shoulda checked it before runnin up.'

'Na, Boss. Ain't that. Would ya just send a couple of mechs out here to trip the Main Landing Gear Safety switches so we can open the canopy?' was my humble response.

The Line Boss laughingly howled back, 'Adams, ya know, that's what ya get for not checkin them fuel gages ya know, hahahaha! Will do, over and out'.

About five minutes later a couple of our own shop buddies pulled up with an NC-5 Jeep at the high-turn run-up site laughing their heads off. They asked us what tripping those switches was worth to us; like maybe taking an extra duty night watch for them? Jayse and I both just said we'd discuss all the details when finally breaking free from our confining situation. Then with great laughter from everyone, the switches were tripped and both of us mindless switchology experts made an escape from our entrapment.

On a grandiose scale with possible 'switchology goofs' costing millions, on 15 November 2007, a brand new Boeing Airbus A340-600 was damaged beyond repair. The 'totally smashing' event happened during ground testing by the delivery crew at the Airbus Facilities at Toulouse Blagnac International Airport in France. As documented by multiple media sources, during an engine test prior to the aircraft's scheduled delivery to Etihad Airways, several safety overrides had supposedly been disabled.

The non-chocked aircraft had accelerated to a speed of thirty-one knots and collided into an inclined concrete exhaust deflection barrier. The collision impact raised the aircraft nose several feet and subsequently cracking airframe structures. The cockpit cabin section broke off, fell over the barrier wall, and onto the ground from a significant height. The aircraft right wing, tail section, and two left engines also impacted other surrounding areas. Ground crash and fire crew services could not shut down one non-damaged engine from running on an accumulated fuel source for nearly seven hours.

Reportedly, nine people on board were injured; four of them seriously (this report varies). The aircraft was stricken from fleet inventory and written off as a loss. What remains of this stricken aircraft is now used at the Cabin Crew Training Facility for the Virgin Atlantic Airways in Crawley, England (just south of London).

There have been rumors of all kinds (and emails galore) sent all over the world about this incident. Tales have been told that the crew had not read the run-up manuals and had no clue about the operating perimeters of the aircraft. Another tale was the take-off warning horn was blaring away in the cockpit because all four engines were at full power. Therefore, the aircraft computers discerned take-off was imminent and flight controls had not been configured correctly. Thus, one of the delivery crew popped a circuit breaker to silence the alarm and placed the aircraft computer into 'thinking' it was in the air. The computers then automatically released all braking power and the aircraft sky-rocketed forward for a lofty trajectory.

Regardless of what really happened (*rumors or reality*), the truth will *NEVER* actually be revealed publically because of extensive embarrassment to more than one agency. However, the result was a shattering $200 million ending to a brand new Boeing Aircraft. In the opinion of this book's author, after serving over thirty-five years in the military (and with over twenty of those years directly related with aviation) *switchology most likely had something to do that aircraft's tragic demise*.

Most likely those in the delivery crew for the aircraft, *should have* been more aware of that particular Boeing Aircraft's operational procedures. That day when running the T-2 Aircraft, I *should have* paid more attention to the aircraft operator's manuals. Shoulda-coulda-wouldas with all its switchology faults of excuses will NEVER take the place of being fully aware of the technical and operator's manuals of the equipment *we are responsible for operating*.

U.S. President Abraham Lincoln is quoted as once saying, 'I will study and get ready and perhaps someday my chance will come.' Therefore, when traveling the pathways of our lives, it can pay great dividends (and even eternal ones) for *knowing what is in the books* which can mend our lives – and *adhering* to their principles. The Holy Bible is one of those *Books of Life Principles,* which can keep us from falling into many incriminating switchology predicaments of life.

Boeing Airbus A340-600 after its 15 November 2007accident at the Toulouse Blagnac Airport

CHAPTER XXI
IDENTIFICATION

"Hey Sarge, so we find this guy and …
he's got no ID. So whata we gonna do?
Ok…guys...Sarge says ta shoot'em"

The U.S. Bureau of Investigation (BOI) was established in 1908. Then in 1935, the agency's name was changed to the U.S. Federal Bureau of Investigation (FBI). By 1946, the FBI had processed 100 million fingerprint cards in a manually maintained system of files. By 1971, twenty-five years later, the FBI had processed 200 million fingerprint cards. The current number of fingerprints on file in the FBI's database is classified. However, according to varied Internet sources, the number is well over 450 million fingerprint files. The FBI's fingerprint files became not only a ready resource for identifying individuals, criminal or not, but also a *very massive one*. As of the writing of the book, the FBI adds 13,000 fingerprints daily to its ever enlarging database.

In 1974, four employees of the UK's Hertfordshire Fingerprint Bureau contacted fingerprint experts throughout the UK. They began an organization of that country's first professional fingerprint organization, the National Society of Fingerprint Officers. The organization initially consisted of only UK experts, but it was quickly expanded to an international scope. Then in 1977, the organization was renamed The Fingerprint Society. With the introduction of a technology called the Automated Fingerprint Identification System (AFIS), their files were divided into computerized criminal files and manually maintained civil files. A few of their manual files became duplicated, but their total records actually represented approximately twenty-five to thirty million criminals, and an unknown number of individuals in their civil files.

In 2005, the International Police's (INTERPOL) Automated Fingerprint Identification System repository exceeded fifty million fingerprints for important international criminal records from 184 member countries. Over 170 countries have 24x7x365 interface ability with INTERPOL's expert fingerprint services.

In 2012, the largest AFIS repository in America began being operated by the U.S. Department of

Homeland Security. It was the Visit Program, contains over 120 million fingerprints, and many of the records are in two-fingerprint format. The two-finger records are non-compliant with FBI and Interpol standards but sufficient for positive identification and valuable for forensics (this is because index fingers and thumbs are the most commonly identified crime scene fingerprints). Since 2007, the Visit Program has been migrating from two flat (not rolled) fingerprints to ten flat fingerprints. A 'Fast Capture' research program will soon enable implementation of a 'ten rolled-print-equivalent fingerprint' recording (within fifteen seconds per person fingerprinted) in future years.

One of the largest ten-print AFIS repositories in America is the FBI's Integrated AFIS (IAFIS) located in Clarksburg, West Virginia. IAFIS has more than sixty million individual computerized fingerprint records (both criminal and civil). Old paper fingerprint cards for the civil files are still manually maintained in a warehouse facility. All military related fingerprint cards received after 19 May 2000, have been, and will be, computerized and can be searched internally by the FBI. Within the Next Generation Identification Program, the FBI may possibly make civil file AFIS searches available to U.S. law enforcement agencies through remote interfaces.

The FBI has plans for eventually expanding their automated identification activities to include other biometrics such as palm, iris, and face. All states of the U.S. and many large cities have their own AFIS databases. Many agencies also store and search biometrics such as iris prints and palm prints. Imprinting interface standards are important to law enforcement agencies as the process enables the sharing of data and reciprocal searches for identifying felons.

In the spring of 2012, the Unique Identification Authority of India (UIAI) is scheduled to possess the world's largest fingerprint multi-modal biometric system – AADHAAR. The AADHAAR term comes from a word meaning 'the foundation' in several Indian languages. AADHAAR is *supposedly* a voluntary program, yet it contains over 200 million fingerprint, face, and iris biometric records. UIAI plans to eventually collect as many as 600 million characteristic records by the end of 2014. With AADHAAR, UIAI has an ambitious goal of eventually providing reliable national identification documents for most of India's *1.2 billion residents*. With a database many times larger than *any other in the world*, AADHAAR's will have the superb ability for worldwide leveraging of fingerprint, face, and iris characteristics in rapid and reliable automated search and identification modes – not just for India, but someday for the entire planet Earth.

Recall the chapter on Military ID Cards? Its dialog went from early clay fingerprints, to ink prints on contracts, to ID cards, to the CAC. The CAC established a 'state of the art' definite *authoritative* process for the use of identity credentials for today and into the world of tomorrow. For security reasons, agencies need identification systems for the integrity of their organizations and safety of their members.

On a different level of identifiable security; when military vessels encounter other vessels (especially in times of conflict), their crews MUST have knowledge of the other vessel status - whether it is ally or enemy. Radar and sonar waves contain anomalies with their reflections in which the equipment operator can sometimes determine if the 'signatures' of those wave natures are friendly or an enemy. In post WWII years, the Identification Friend or Foe (IFF) electronics system was invented and implemented in all weapons systems. The weapons transponder broadcasts a secure code (changed daily) and communicates with towers and other vessels. If the transponder code does not match for that particular day, then the vessel is considered 'Foe' (or hostile), and regarded as an enemy.

Thus, even radar, sonar, and IFF systems are used as positive protection recognition equipment for agency level secure identifiers.

On a personal level, everyone wants to *identify* with some cause, some event, something, and especially that one special 'certain someone' for making life worthwhile. However, with that concept on an agency level, some identification measures could be taken too far and out of control. In contemplating the need for identity, use of ID cards, and the ever expanding databases countries have use of with their citizen's personal 'identity' information – consider what would happen if someone could usurp governments and assume control of all databases.

Such usurping actions might be viewed as comparable to being like a global coup with just one person in charge – on a global scale.

This concept may sound far-fetched, but during the 2009 G8 Meetings, Pope Benedict released a new papal letter stating "there is urgent need of a true world political authority." His statement was further expounded as the mentioned authority needed to (1) manage the global economy, (2) revive economies hit by the crisis, (3) avoid any deterioration of the present crisis and the greater imbalances that would result, (4) bring about integral and timely disarmament, food security and peace, and (5) guarantee the protection of the environment and to regulate migration.

Therefore, even *organized religion* has come to the point of seriously considering all the world really needs is a one-world leader; one to settle all disputes and lead the world to peace. Not only has the Pope mentioned this concept, but it has been talked about by politicians for decades.

This 'far-fetched' concept is really not a new one; it is mentioned in the Holy Bible many times. However, with the advent of global identification and great new technologies, such a concept becomes more conceivable as the world rushes madly into the 21st century. A couple of my other books have an entry titled, '*The Voice of God.*' The point behind that entry has some relevance for this entry of *'Identification'* and its entry is the next one.

CHAPTER XXII
THE VOICE OF GOD

"Hey, did you just hear that strange voice?"
"No, I'm kinda busy right now shootin folk!"

'But you really don't remember, was it something that they said? Are the voices in your head calling, Gloria?' Recall those lyrics from the song *"Gloria"*? The song went to number two on the U.S. Billboard Top 40 Charts in 1982. Italian pop singer Umberto Tozzi originally recorded the song as a ballad in 1974, but Laura Branigan sang its most popular, best-known version.

Thinking about 'voices in our heads,' how do we know when the Lord God impresses us with His Words or His thoughts? How do we know that He has actually 'spoken' to us or 'spoken' into our spirit from His Spirit? The answer to that towering question is actually a personal one. The Lord God connects with each individual in a personal way. The way He communicates with me may not be exactly the same as the way He communicates with others. Yes, God does speak audibly, but it is rare. In the Holy Bible were Prophets in antiquity who were devoted to following God and communing with Him. Because there was little to no written Word of God in those days, God chose the use of Prophets as His mouthpiece. The Bible also records where God used angels as His messengers for carrying His Words.

How then does an individual know the voice of God impressed into their heart, mind, or spirit? The follower of God through Christ must follow in a close relationship. Best friends sometimes know each other so well they know what each other will say next because they know how they think. This is because they are in each other's presence so much of the time, and that familiarity forms some bonding of shared mind and spirit characteristics. That principle works the same with the Lord God and us. Then when God speaks, or impresses something of Himself into us, we sense it, and know within us that it was from Him.

God spoke to the Prophet Elijah once when Elijah was going through a difficult time with being threatened by Jezebel (reference Chapter 19 of I Kings). God told Elijah that after all the winds came and tore into the mountains, the rocks had broken, an earthquake came, and a fire erupted, He was not in those things. After all those tumultuous things, He was in the still, small voice, which followed them. God's voice and directions are usually not the loudest and most salient things we hear and/or observe. He is this way because He wants our spirit to be in contact on a regular basis with Him so that regular familiarity identifies Himself to us through all the clutter when He speaks.

Consider again 'all those voices in your head' from the lyrics of "Gloria." What if there were ways to project voices into the heads of people? What if the technology existed, or was soon to exist, which could project thoughts into the minds of individuals? How would individuals know if thoughts were actually their own originated thoughts or implanted thoughts? Considering this further, how would individuals know the voice of God or impressions from God?

Consider a military weapon called 'The Voice of God' or 'The Voice of Allah.' Conducting Internet searches on this technology will reveal very surprising results. Those results are that the weapon does exist, does not exist, is in developmental stages, and/or is very speculative. The results can be found scattered in editorials within *The Washington Post* to *The Star*. The Voice of God Weapon is a device apparently being developed by the military and operates at a distance, delivering a message in which only a single person can hear. The messages are in effect that of 'Stop the bloodshed,' 'Surrender,' or some other command to the benefit of the hearer's opponent.

During the development of this book, there have been recent technical workshops reporting 'The Voice of God Weapon' has successfully been field-tested. Results indicated that one insurgent in each subject group would begin looking in all directions and then engage controversial conversations with his companions – those who did not hear the 'implanted' message. The purpose of this weapon is the disruption of thought processes of the opponent. The end result is confusion within the opposing troops and beginning the breakdown of communication within the chain of command, thus, rendering opponents useless and defenseless.

As with all weapons used for defensive purposes, they can also be used for offensive strategies as well. This same weapon in the wrong hands could have devastating potential, and one of those is with terrorism. Adequately funded fanatical terrorist factions could use this weapon to their advantage against good peoples and nations with well-meaning societies. Just ask any good soldier, sailor, or airman from field, and it will be verified that any weapon has good and evil potential. The end result of any weapon is the intended goal.

Having considered intended weapon goals, look further into the Voice of God weapon falling into the hands of a one-world leader. With the advent of global transportation and communication in ever-reducing time frames, the world is becoming 'smaller' socially. This current world is shaping up politically, economically, socially, and religiously, for one government to solve all of its global problems. Several hundreds of books have been written about the possibilities of one-world governments and leaders. With nightmarish problems impacting the global community, these problems become more and more of world concerns by world citizens. Therefore, one leader could arise and make a difference among all of them – in theory.

What if a leader in the world's celebrity political scene did arise in popularity? What if that one leader possessed the charisma to sweep the world beneath their feet? What if that one leader was given leadership authority over all governments of the world? Then, what if that one-world leader became corrupt and perfected the Voice of God Weapon for a broader-scale application? The regime of this leader could use it to program the citizens of the entire planet Earth. Programming could exist at voting stations, licensing bureaus, application booths, advertisement stands, shopping centers, et cetera. The list could literally go on and on. The most obvious places for programming would be the places of employment, especially federal and state governments. Secondary places would be through the media of radio and television (when the technology is enhanced).

Now, enter the Antichrist. The Holy Bible also calls him the Beast in several passages of Scripture. This man is the epitome of evil because he is actually satan incarnate. He will imitate everything that Christ did in order to confuse everyone who believes on the one true living Lord God Almighty. For many years, it has been believed the Antichrist will have the ability for instituting a global numbering system for tracking everyone. Numbers issued by the Antichrist regime will be indelibly imprinted on the right hands or foreheads of every person and will track the peoples of earth (reference Revelation 13:17-18 and 14:11).

The Antichrist will also have the authority for enforcing all citizens of planet Earth to worship the Image of the Beast. This is an image of himself, and all who do not worship this image will be put to death. This is found in Revelation 13:15, *"He [Antichrist] was granted power [by satan] to give breath to the image of the beast* [Antichrist], *that the image of the beast should both speak and cause as many as would not worship the image of the beast to be killed."*

What is to say that The Voice of God Weapon technology could, or could not, be modified for use in brainwashing techniques of mental reprogramming? If the one-world leader in charge wants it done, who is to stop the motion, especially if that entity has god like powers? What if The Voice of God Weapon technology were used to enforce submission to the new one-world regime and especially its leader – by way of worship? And that worship just might be at the television, where a hologram of the Beast could someday be projected.

These ideas may be a little far-fetched, but what if I am not wrong? What then? How will you know the Real Voice of God versus The Voice of God Weapon technology implanting something in your head? These concepts ought to be seriously considered, especially since The Voice of God Weapon technology is a reality, despotism in the world is a reality, and there is a real devil out there in reality. So then, how would you discern the difference between the real voice of the Lord God and the false voice of god technology?

Christ Jesus spoke much in Chapter 10 of John about His people knowing and hearing His voice. Then He said in John 10:27-30, *"My sheep [His people] hear My voice, and I know them, and they follow Me. And I give them eternal life, and they shall never perish; neither shall anyone snatch them out of My hand. My Father, who has given them to Me, is greater than all; and no one is able to snatch them out of My Father's hand. I and My Father are one."* Christ was stating that if we know Him personally, we can hear His voice when He projects it within us. Whether His voice is in a dream, a vision, an impression, a sensation, a thought, or an actual voice – we will know it came from Him because we know Him.

Then later while Jesus was on Trial and talking to Pilate, He said, *"...Everyone who is of the truth hears My voice"* (John 18:37). One must seriously consider something here with this passage of Scripture. When coming to God through Christ Jesus, repentance of sins is required. This is being sorry for the wrongs committed and wanting to turn away from them. Then with believing Christ's sacrifice on the cross as payment for sins and accepting it, a regeneration of the inner spirit begins with the rebirth of the soul. That regeneration process is a continual process until death comes.

That regeneration process can be fast, slow, shallow, and/or deep. It depends on the interest of the individual. The deeper the interest will usually generate a faster process. Additionally, the deeper process will bring the individual into a very close personal relationship with the Lord God. The deeper the relationship, the more truth and honesty is desired and earnestly sought after. In fact, the deeper the relationship, the more truth and honesty are hungered for. The more hunger for truth and the more in communication with God (talking with Him), the clearer becomes the sound of His voice. Thus, the dedicated Christian will have no difficulty whatsoever distinguishing the Lord God's voice from any confusing voice of god weapon technology.

Discerning the Lord God's voice, at such a time of this twenty-first century with all its technologies, is truly a personal impression from God; and God knows how to imprint Himself into each and every one of us. We need to earnestly seek the encoding processes. We also need living a life close to God, stay connected to His Spirit, and remember Isaiah 54:17, *"No weapon formed against you shall proper...This is the heritage of the servants of the LORD..."*

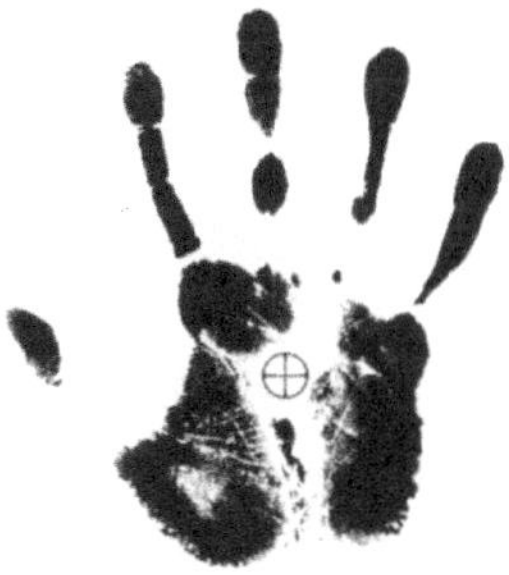

Beware of world leaders insistent on having '*their identifying marks*' inscribed in palms of hands and forehands. According to the Holy Bible, their regimes will lead into totalitarianism; and will also be the voice of one leading to eternal damnation.

CHAPTER XXIII
WILD BLUE YONDER

"So, ya wanna hitch a ride do ya? Well take a number. I got bombs to drop!"

'Off we go into the wild blue yonder, Climbing high into the sun; Here they come zooming to meet our thunder, At 'em boys, Give 'er the gun!' Those were the first few words of "*The U.S. Air Force Song*" and the official song of the United States Air Force.

In 1937, the Assistant Chief of the U.S. Army Air Corps, Brigadier General Henry Arnold, persuaded the Chief of the U.S. Army Air Corps, Major General

Oscar Westover, that the Army Air Corps needed an official song. The song should reflect their unique identity in the same manner as that of the Army, Navy, and Marine Corps. He further proposed a song competition with a prize for the winner. However, the Army Air Corps did not have control of its budget and could not award a prize.

In April 1938, Bernarr MacFadden, publisher of *Liberty* magazine, assisted by offering a prize of $1,000 to the winning composer. There were stipulations, though; the song must be of simple harmonic structure, within the limits of an untrained voice, and its beat must be in the marching tempo of a military pattern. Over 700 compositions were received and evaluated by a volunteer committee of senior Army Air Corps wives having musical backgrounds. July 1939 was the dead line given by the committee for choosing a final selection. However, in spite of the great number of entries, information eventually spread that the committee had not found any songs particularly interesting. Then in 1938, the Chief of the Army Air Corps solicited inquiries from professional composers and commercial publishers.

In 1939, just two days before the deadline, Mr. Robert Crawford, a professional musician, music instructor, aviation enthusiast, and a former Army Air Corps Pilot, and also who was widely known as the 'Flying Baritone,' personally delivered a sound recording of his entry. The song caught the committee's attention and they called upon Mr. Rudolph Ganz; a guest conductor of the National Symphony Orchestra and also a consultant for the committee. The committee wanted his evaluation of

Mr. Crawford's entry. Mr. Ganz was in favor of the song and then the committee immediately, enthusiastically, and unanimously approved Mr. Crawford's entry for being the competition winner.

Mr. Crawford had originally titled his song, *"What Do You think of the Air Corps Now?"* but it was soon after officially changed to *"The Army Air Corps."* Mr. Crawford publicly sang the song for the first time over national radio from the 1939 National Air Races. In 1947, when the Army Air Corps became the U.S. Air Force, the words 'U.S. Air Force' in the title and lyrics replaced the original words of 'Army Air Corps'. On 27 September 1979, General Lew Allen, Jr., Chief of Staff of the Air Force, adopted the song as the official song for the U.S. Air Force.

Inspiration from that song has prompted many into bravery from the valley of defeat. Its words have elevated the human spirit in pressing toward winning against surmounting odds. Yet, there are times in life when daring decisions will cast long shadows of horrendous mistakes down the pathways of broken dreams – all because of *pride and misinformation*. Conducting a little research, and in some cases, a more in-depth research, facts can be found which tend toward proving that *pride and misinformation,* to some degree, caused (or may have led) to the following very costly situations both military and civilian (military designated with an 'M' and civilian with a 'C'):

General Meade and Indecision (M): In July 1863, it was the American Civil War and the Army of Northern Virginia was done. Three battering and bloody days at Gettysburg had reduced the Confederate soldiers to just a skeleton of their former strength. With all his resources and reserves spent, General Robert E. Lee had assembled his badly beaten forces and was attempting to make it back to Virginia. In his return march, rains had greatly swollen the Potomac River banks, and on his flanks were persistent Union Cavalry troops.

The rain soaked roads had become a quagmire of mud. In that defeat march, the stage was set for the final crushing blow to be delivered by the Union Army of the Potomac (and, it had several fresh reserves). All that needed to be done was sweep down on the defeated soldiers of grey like a blue wave. The Army of Northern Virginia would be totally crushed and the American Civil War would then be over. All that remained was the order from General George G. Meade of the Union Army to attack.

But that most decisive order never came.

For reasons unclear to history, General Meade became reluctant to follow General Lee. Instead, he collected his forces in strength and waited. No one was ever quite sure what his thoughts were or what he was waiting for. However, when President Abraham Lincoln discovered that General Meade had *literally allowed the end of the American Civil War to slip through his grasp*, ole Honest Abe became infuriated.

For the most part, it was General Meade's indecision which resulted in General Ulysses S. Grant being called east from Vicksburg and placed in command of the Union Army of the Potomac. Had

General Meade attacked the defeated and retreating Confederate Army at that one most opportune moment, the American Civil War probably would NOT have lingered on for nearly *TWO more years.* No one will ever know the thousands of lives (Union and Confederate) that could have been spared with just that *ONE decisive strategic military move.*

Had that *one decision* been made, the Post Civil War Reconstruction Period would most certainly have been very much different – and the United States of America may have taken an entirely different historical course than it has.

Colonel Custer and Guns (M): In 1876, the relatively new patented Gatling Gun was still in its developmental stages as an early type of rapid-firing gun. It had completed its trials, and Lieutenant Colonel George A. Custer had two to four of those infamous rapid-firing guns along with abundant ammunition in his command. In fact, they were available for his use when setting out toward a small Indian village on the bank of the Little Bighorn River.

Colonel Custer was going to settle some reported unrest among the Indians and settlers. On their bivouac, though, Colonel Custer chose *NOT* to bring his Gatling Guns and ammunition. His reasoning for not bringing them, *and possibly using them*, was that they would impede his march and hamper his mobility. More importantly, though, Colonel Custer is also said *to have believed* that the use of so devastating a weapon would '*cause him to lose face with the Indians.*' Considering known reports of Colonel Custer's vanity, actions as this are not hard to believe.

Over 250 soldiers died on a Montana hill on 25 June 1876, under the leadership of a Colonel who made one decision which no doubt had it been made the other way would have saved most of their lives - both U.S. Army soldiers and American Indians. Just the presence of the Gatling Gun that day, would probably have been enough for instilling serious thinking and then some subsequent peace talks.

Admiral Halsey and Typhoons (M): It was WWII and Admiral William 'Bull' Haley's Task Force 38 (TF 38) had been operating about 300 miles east of Luzon in the Philippine Sea. Naval aircraft had been conducting air raids against Japanese airfields in the Philippines. The fleet was attempting to refuel its ships; but, as the weather worsened it became increasingly difficult to refuel, and all refueling attempts had to be discontinued. Despite warning signs of worsening storm conditions, the ships of the fleet remained in their battle stations.

Making matters worse, reports for the location and direction of an approaching typhoon were not accurate. Then, on 17 December 1944, Admiral Halsey sailed the U.S. Third Fleet right into the heart of Typhoon Cobra. Nine warships were severely damaged, three warships capsized and sank, over one hundred aircraft were wrecked or washed overboard, and 1,450 lives were lost while encountering Typhoon Cobra.

Whether because of mistaken reports, or the drive for combat in time of battle toward the ending of WWII, this event was very costly. It was just as costly as losing a battle.

President Truman and Korea (M): Wars are best run by the military professionals and not micro-managing politicians. On 25 June 1950, the North Korean People's Army invaded South Korea; the resulting outbreak of conflicts began the Korean War. By early 1951 the war became a fierce stalemate near the 38th parallel with United Nations (UN) and U.S. casualties becoming heavy. President Harry S. Truman rejected General Douglas MacArthur's request to attack Chinese supply bases north of Yalu. General MacArthur, however, promoted his plan to the Republican House leader, who leaked those plans to the press. President Truman then became solemnly concerned the war would further escalate and the Soviet Union could possibly be drawn into its conflict (it was already supplying weapons and aircraft to North Korea). Therefore, on 11 April 1951, President Truman relived General MacArthur from all his commands in Korea and Japan.

At the general's position and status, that 'political' move was a firing and effectively resulted in removing the general from military service. That action by President Truman began his demise. His ratings steadily plummeted and many called for his impeachment. General MacArthur was perhaps one of the greatest generals of the 20th century, and the entire world knew it. Had the firing of General MacArthur never happened, and the general's plans been implemented, the Korean War may have ended long before it did on 27 July 1953.

Just think; in those possible remaining two years of the war, millions of dollars that were expended could have been saved. Just think; in those possible remaining two years of war, how many lives could have been spared from all sides? But, they were not – and all because of one decision. And why was that one decision made? Only President Truman knows the truthful answer to that question.

President Johnson and Vietnam (M): Wars are best run by the military professionals and not micro-managing politicians (Part II). During the American involvement in Vietnam with President Lyndon B. Johnson, he became a micro-manager of detailed war events and monitored them too closely. That form of management changed the role of 'American Advisors' in a small insurgency into becoming an all out 'American Police Action.' Thus, nearly 60,000 American military lives would be claimed with that micro-management practice before the conflict ended two presidents later.

Just think; had one man not wanted to micro-manage his country's involvement in war, millions of dollars that were expended could have been saved. Just think; had one man not wanted to micro-manage his country's involvement in war, how many lives could have spared been from all sides? But, they were not – and all because of one micro-management style.

And why was that one decision made? Only President Johnson knows the truthful answer to that question.

Afghanistan Blunders (M): For centuries countries outside of Afghanistan – from the Indian Mughals, to the British Empire, and the Islamic fundamentalists – have all tried imposing their will upon the peoples of Afghanistan. They have all failed; and as a result, the Afghan people have become a very resilient populace who are experts in guerilla warfare. The Soviet Union either ignored, or totally forgot, all their over 1200 years of important documented history of conflict; for in 1979 they moved on Afghanistan.

The Soviets deployed massive numbers of troops and weaponry for the purpose of establishing their communist regime in Kabul as the beginning of their takeover. What followed was a ten-year blood bath of death in the rocks in Afghanistan. The Soviets suffered over 50,000 documented casualties, over 15,000 KIA, and a cost of billions of Russian Rubles. In return, the Russians obtained *nothing.* For the Afghans, though, their country was left devastated and ripe for a fanatic religious group called the Taliban to take it over.

Since the Soviets departed Afghanistan, the Taliban grew in power and have become a serious threat to the world through proactive terrorism.

RMS *Titanic* (C): The sinking and loss of the RMS *Titanic* in 1912 was a loss of over $168 Million. Most everyone thought the ship was unsinkable. Some even thought (and openly stated) that God could not sink it. While in route, the *Titanic* received a series of warnings from other ships of drifting ice in the area. Yet, the ship continued steaming ahead at full speed. Captain Smith himself had declared that he could not 'imagine any condition which would cause a ship to founder. Modern shipbuilding has gone beyond that.'

Jonestown, Guyana (C): The mass suicide of 909 religious cult temple members in Jonestown, Guyana, in 1978, was a prime example of how failure to accept the truth will lead to deception – and sometimes on a large scale.

Space Shuttle *Challenger* (C): The explosion and loss of the Space Shuttle *Challenger* in 1986 was estimated at a loss of over $1 Billion. The mission had a known 'O-ring' problem at least several months prior to the launch of the *Challenger*. However, the mission goal was to stay on schedule and the problem was down-played as a low-risk situation. Had more credence been given to the O-ring situation, this disaster would most likely never have happened.

Chernobyl Power Plant (C): The nuclear accident at the Ukraine Chernobyl Power Plant in 1986 caused the deaths of 125,000 people and was a loss of over $200 Billion. Fifty percent of the area is still, in some way, contaminated. Certain operators at the Chernobyl plant began conducting unauthorized experiments. They deliberately circumvented (bypassed) safety systems in order to learn more about the plant's operation. Well, guess they found out what it would do. 'Safety First' becomes a *very expensive slogan* if its words are ignored and select steps are bypassed.

The *Exxon Valdez* Oil Spill (C): In 1989, the *Exxon Valdez* oil tanker ship struck an Alaskan reef and spilled 10.8 million gallons of oil into the ocean. The oil spill loss has been estimated at over $2.5 Billion with an unknown ecological toll. Multiple factors were found to be blamed for the oil spill. The primary factor *appears* to be that the shipping company had failed in providing adequate crew rest. *However*, that does not explain another report; one which indicates the ship's RAYCAS (Raytheon Collision Avoidance System) was not operational, and had not been operational for several months before the accident.

Waco Texas Seige (C): The deaths of 86 people occurred during the Waco Seige in Waco, Texas, in 1993, is another prime example of how failure to accept the truth will lead to deception - and sometimes on a large scale.

Boeing Airbus A340-600 (C): The injuries of 9 personnel and total loss of a brand new Boeing Airbus A340-600 in 2007 came to a loss of over $200 Million. When in doubt about a machine's operation, *read the operator's manual!*

Metrolink Commuter Train (C): The deaths of 25 people occurred when a Metrolink Commuter Train crashed head-on into a Union Pacific Freight Train in Los Angeles, California, in 2008. The loss has been estimated at $500 Million (so far). The lesson here is DO NOT text while driving; *especially trains!*

Deepwater Horizon Oil Spill (C): The deaths of 11 men and the injury of 17 others resulted with the Deepwater Horizon (BP) oil spill in 2010. This oil spill became the largest in history up to that time. By the time the gushing wellhead was capped, approximately *five million barrels* of crude oil spilled into the Gulf of Mexico. The spill caused extensive damage to marine and wildlife habitats, as well as the area's fishing and tourism industries. Loss for this accident is estimated at $40.5 Billion (so far). Reports indicated that less expensive operations were being implemented. At the same time, official communications seemed becoming an issue between the involved parties with the cost reduction applications. These measures led in triggering the explosion and the ensuing oil spill.

(All loss figures for the before mentioned examples were in year 2011 estimates)

Pride can be good, not-so-good, and totally destructive in one's life. The attribute depends on the *specific application* in which it is utilized. Pride in one's work and achievements, to a *healthy degree*, will act as measurement guidelines for future goals. Pride in one's appearance and self worth, to a *healthy degree*, will act as a standard against ridicule, and as a generator for self improvement. However, *excessive pride* will entertain and embrace misinformation. Why? Because it loves greed, and misinformation has many ways of bringing dangerous short cuts into any scenario, gaining wealth quickly, while at the same time making these things irresistibly attractive. Pride, mixed with misinformation, is very deceptive; and if its wrongfully-rooted insidious tentacles are not guarded against, they will go *beyond* the point of being blinded by them.

Before the U.S. entered WWII, an American was serving in the British Royal Air Force – Pilot Officer John Magee, Jr. On 18 August 1941, Pilot Officer Magee had flown up to 33,000 feet in a British Spitfire Mark I Aircraft; it was his seventh flight in the plane. As he was spiraling and climbing upward, he was struck with the inspiration for a poem – *To Touch the Face of God*. Later that day after landing, he completed the poem. Its first and last lines are:

> 'Oh! I have slipped the surly bonds of earth...put out my hand and touched the face of God.'

Pilot Officer Magee enclosed the completed poem on the back of a letter to his parents. His father had it reprinted in church publications. The poem became more widely known as *High Flight* and is known by most aviators and aviation enthusiasts around the world. The manuscript copy of the poem still remains at the U.S. Library of Congress. Pilot Officer Magee was killed in a mid-air collision a few months later 11 December 1941. He was only age nineteen.

Unlike Pilot Officer Magee's desire to reach out and gracefully touch the 'Face of God,' far too often in our rebellion and just plain wanting to do things our own ways, inside we feel somewhat like him high in flight. In our running we have escaped the surly bonds of things which have held us for so long. In some cases, higher and higher we propel until our own pride interjects that we are self sufficient and we do not need God. Some may even raise fists at God because they have escalated themselves to a lofty position. Metaphorically, off we go into the 'wild blue yonder' aimlessly at life doing whatever we want because we 'know' we are the masters of our own fate.

Yet, without God's graces (in the spiritual realm), figuratively speaking, the gravity force of nature *always* finds us and we will come crashing down – sooner or later. And, hopefully, we will not collide in mid-air with a friend or foe, and become a total loss to everyone and ourselves.

What if the things we have been told about God are wrong? What if the information we have heard and read is corrupted? It just might be that God is not who we think He is. Each and every one of us needs to be seeking God for ourselves, and asking of Him who He really is. Others may say God does not exist, but where is the actual proof? Others may state that God is a mean Deity, but is their opinion actually out of bias from their rearing and peers?

High flight can accelerate life to the realm of heaven, or, high flight can degenerate life into the depths of evil. Throughout life, God can be our Celestial Navigator, or, the devil can be our deviant deceiver.

High flight will always make a choice – the high road or the low road. The end of that road we choose depends of the navigational aids of choice. For the best outcome with a life of high flight, be sure to read the next entry – *What If?*

CHAPTER XXIV
WHAT IF?

The prefacing material for this entry's beginning is information in other books about God delivering me from a near-fatal head injury and being lost at sea. That information also included visions of skull faces and a Grim Reaper coming to me before the head injury incident. Also included was a dream my mother had about me dying and a vision a friend had of me being in great need. And so, follows is the short story of *What If?*

With my stories, I have given credit to the Lord God for delivering me. Some may say all these things were just chance with the randomness of the factors of life. How does one explain the dream of my mother the night before my head injury? How does one explain the vision coming to my friend while driving the morning of my head injury? How does one explain those skull faces and the grim reaper coming to me just a few months before my traumatic head injury?

The mystic would not discount those strange events as mere chance – and neither should the agnostic or the Christian. Within those events are some elements of truth, for those dreams and visions foretold a later reality. For many years I have been a Christian; I've not been perfect, but as a Christian I have been forgiven of my imperfections. All through my years as a Christian I have seen God prove Himself to me that He is real. The stories in this book are but a few of His proofs given for me. My other books contain many more stories of why I believe in the Lord God.

For those now reading this book, who may not believe in God, I put forth this question to you – What if it's true? What if there really is a God?

That 'truth' opens up a whole realm of reality both for this world and the next; and thus, interweaving the physical and spiritual dimensions. If there is a God, then is that God interested in us – humanity? If there is, and if He is, then to what extent would that God go toward assisting us along our difficult pathways in life?

There are many answers to that question, and this book addresses two of those. One is briefly discussing a fundamental concept of faith in God and is in this chapter. Another has been revealed with three of my strange life encounters; that God is definitely interested in our sojourning of this road we call life. He is interested enough to inform us, and others, of what may happen to us. He is also compassionate enough to provide rescue for us when there is no hope at all. And now, for the faith factor, examine a few interesting concepts.

When finding one's way (especially when lost in any situation) it is very important to know *who* to believe, *what* to believe in, and *when* to make the right decisions. When traveling and becoming lost, an incorrect map will not be of any benefit whatsoever. Having the map of a location other than the area where we are located is totally useless. Likewise, what about a belief in God? Some believe in God, and some do not. Some believe that because we cannot see God, He does not exist. Some believe that a good God would not let bad things happen at all.

The wind blows, and we do not see it; but we *feel* it and we *know* it is there. Our eyes cannot 'see' the realm for the energy wavelengths of the winds. Yet, our skin can detect the pressure of those wavelengths of energy. One of our senses cannot detect the presence of the winds, *yet* another one of the senses empathically detects the *evidence* of its presence. Such is the same with God. We may not be enabled to see Him; however, the evidence of His existence is present, and has been around for eons of time. The question really is then, 'Do we have the *faith to believe* the evidence He has presented?'

This is a fallen world, the dominion of satan and, therefore, evil things *will* happen – <u>***even***</u> to good people. Such does not negate the existence of God, but actually proves it. All through the events with the existence of a fallen world are scattered occurrences of God's intervention for extraction, improvement, redirection, comfort, forgiveness, etc. God has chosen to prove Himself in the evil corridors of a fallen world. It is in those avenues of darkness that His Light shines on the pathways for life's traveler to navigate correctly and efficiently.

It is up to each individual life traveler to access the Light given them, to determine if there is a God, and what He means to them personally. It is with that Light whereby the life traveler sojourns a safe navigation home to Heaven after this life. Without embracing that Light, the life traveler will grope in darkness and never find God.

There is constant rhetoric about whether the Holy Bible is truth or just fables. Is it a work of fiction or non-fiction? This stuff has been around in various forms since the Garden of Eden, and it will not cease until God brings an end to the age of man. It is appalling to discover the high numbers of people in this world who believe the Bible is just a bunch of hand-me-down stories, only good for soothing the soul at bedtime, and not actually correcting its wrongs.

Jewish candidates of antiquity entering the Rabbi ranks were required to memorize the entire Torah, the first five books of the Bible. The societies of today's world fail in comprehending that many Jewish Rabbis know all of the Talmud's 2,700 plus pages perfectly by heart; word for word. The Talmud is a compilation of legal opinions and debates and also contains the first five books of the Bible. Many of today's Jewish Rabbis are very unimpressed by the computerization of the Talmud, for it means very little to have it in any computer memory drive, when it can be embedded within their own living memory banks. This memorization task is word for word, and is taught and administered by their Rabbis. It is totally illogical for even considering any errors in the Torah when every Rabbi in the nation had (and still has) the Torah completely and flawlessly memorized.

As centuries passed, more tales of errors in the hand-me-down stories of the Bible became commonplace. Then came the Dead Sea Scrolls, and their discovery cast away many shadows of Biblical errors. The Dead Sea Scrolls proved the accuracy of the transmission of the Bible. Those scrolls were discovered in Qumran Cave One near the Dead Sea in 1947. There are in existence some Old Testament manuscripts dating back to about 150 BC. Comparing those manuscripts with the Dead Sea Scrolls proved to be basically the same, with only minor variations due to penmanship slips and spelling variants. The fact that those manuscripts were distanced by nearly two thousand years and were basically the same is an indicator of the fantastic accuracy for the Bible's manuscript-copying transcription efforts from antiquity.

An entire copy of the Book of Isaiah was also discovered at Qumran. Even though the copy of Isaiah discovered at Qumran Cave was a thousand years earlier than the oldest dated manuscript previously known (980 AD), it proved ninety-five percent word-for-word identical with the standard Hebrew Bible text. The five percent variation was, again, mainly pen slips and variations in spelling.

Manuscript discoveries like the Dead Sea Scrolls provide undeniable evidence that today's Bible Scriptures are (for practical purposes) exactly the same as when originally inspired from God and recorded by His Prophets. The Dead Sea Scrolls prove that the scribes (copyists) of Biblical manuscripts took great care in going about their work. Those ancient scribes knew they were duplicating God's Word. Therefore, they treated it with great respect and went to all extents possible in preventing error from entering their literary works. Those scribes carefully counted every line, word, syllable, and letter for complete accuracy. They realized that someday they would give an account at God's Judgment for their penmanship.

As for just the soothing, feel-good, hand-me-down, stories of the Bible; consider just two of those stories. One is Jonah in the belly of the fish, and the other is Noah with the big ole boat. Most of the world scoffs at these stories being so ridiculously untrue; better rethink them again. Jesus referred to both of those men as truthful events when He spoke in Matthew 12:40 and 24:37. Therefore, if those events were not factually true, why would Jesus reference them as true? There are stories of Jonah in other cultures. His story is also written in the Koran (Qu'ran), and the Bahá'í faith (Bahá'í is a monotheistic religion founded by Bahá'u'lláh in nineteenth-century Persia). Therefore, in that whale of a story, there must actually be something to it for other ancient cultures having recorded it. The most common story of Jonah is written in the Jewish Bible, from which Christianity obtained it.

Jonah, though, is not the only person who has been purportedly swallowed by a great fish, and lived. There are also legends about men being swallowed by whales. One story is about James Barley, who in 1891, while sailing aboard the whaling ship *Star of The East*; reportedly fell overboard; was supposedly swallowed alive by a whale; was recovered later; and lived. Another story is of a Marshall Jenkins, who in 1771, while sailing aboard the whaling ship *Edgartown*, came into contact with a whale. The whale broke one of the smaller whaling boats in two, took Jenkins into its mouth, and submerged with him. Later returning to the surface, the whale vomited Jenkins onto the wreckage of the broken boat. Jenkins was very bruised but not seriously injured. Those big fish stories cannot be substantiated as factually truthful. Why? One major fault is that depending on where one obtains research information, much of the resulting data is very conflicting. Some data even states the ship's logs do not even list those sailors as having served with them.

There are two very interesting facts for consideration, though, about the Jonah Story. One comes from Berosus, a 4th century BC Babylonian historian, and the other appears much later in history. Berosus recorded many traditions of the Assyrians where Jonah was directed by God to go minister. One of those Assyrian traditions was the worship of a fish-man, who came up out of the sea. During the time of Jonah, the people of Nineveh believed in a divinity sending them messages by a person rising out of the sea. They believed he would be part fish and part man. Oh, you can be assured that when Jonah 'arrived in style' being vomited up on the beach from a great fish, they undoubtedly were very receptive to anything Jonah had to say. God's timing and ways are perfect, are they not?

The other interesting fact is the preservation and discovery of the name 'Yunas' in the ruins of ancient Nineveh. In 1847 the British actually began limited excavation of Mosul, or ancient Nineveh, and intermittently continued the task for decades. Nineveh had seven entry gates within its wall; one name was known for centuries as 'Neby Yunas.' When modern archeologists unearthed a mound at that gate, beneath it was found the ruined palaces of Nineveh's kings. Also was found the tomb of Neby Yunus, which means 'Prophet Jonah' (Yunus and Yunas are variations in the spelling of Jonah), thus, confirming the historicity of the Jonah Story.

Whichever way God transported Jonah to Nineveh, it for sure had something to do with a great fish. If Jonah were swallowed by a great fish (which I believe), then God could have temporarily shut down the gastric system of the fish (which has for centuries given skeptics problems with this story). God could have also extended the available oxygen Jonah needed for his stay in that fish. Who knows, God could have even lowered Jonah's metabolic system similar to a hibernation-like state.

As for that ridiculous big ole boat that Noah could not possibly build (and 'load all them' animals on it), there is an Internet website which shows some interesting photos of the remains of Noah's Ark. Go to the specialtyinterests.net website and search for the remains of Noah's Ark on Mount Ararat in Turkey. The resulting data and photographs are very surprising. May I strongly suggest readers take a look, and read over the recorded documentaries? Whichever way God chose in Noah building that Ark, God made sure it was big enough and strong enough for carrying ***all*** the animals God wanted on it ***at the time***. Consider all the differing races and types of people that came from just those eight people of Noah's family. That same concept can also be applied to the animals as well.

Consider that Jesus came from God, and God is truth, and God cannot lie. Therefore, Jesus cannot lie. As a result, the stories of Jonah and Noah are, thus, validated as true by the words of Jesus, the Son of God, aka Mr. Universe of Truth. The Dead Sea Scrolls were actually God Proofs of His existence, just waiting for modern day archeologists to discover them for the skeptics. The finds and photographs of Jonah's Tomb, the Gate in Nineveh, and Noah's Ark in Turkey are God's imprints for the skeptics over time. It has been said that seeing is believing, yet with some things, some individuals still cannot believe, even with a print from God.

In further pondering the question of 'What If?' there is a fact which needs serious consideration. An interesting fact for increasing faith that there is a God is not found in the Bible, but actually within mathematical science. Most everyone in this age has heard the phrase 'Do the math.' It implies the application of sound rationale to a problem and arriving at a logical conclusion. Following that train of thought, there is an actual mathematical probability model of Christ Jesus being who He claimed Himself to be. Christ Jesus claimed to be the long prophesied Jewish Messiah, the Son of the living God.

Dr. Pete Stoner was a Christian Mathematical Scientist in the 1950s and 1960s. He was the

Department Chairman of both Mathematics and Astronomy at Pasadena City College until 1953; Chairman of the Science Division, Westmont College, from 1953-57; Professor Emeritus of Science, Westmont College; and Professor Emeritus of Mathematics and Astronomy, Pasadena City College. Dr. Stoner wanted to look at the statistical probability of Jesus being who He said He was.

Dr. Stoner once took the elements of sixty prophesies within over three hundred references contained in the Old Testament declaring the coming of, character of, and the life and death events of the Lord Jesus Christ, and programmed them into his computer. In the magazine *Science Speaks* during the mid 1960s, Dr. Stoner revealed those resultant findings. He wrote that using the laws of mathematical probabilities, there exists a 10^{157} possibility that someone could fulfill all of those prophesies in an average lifespan of about seventy years. The algebraic number 10^{157} is 1 with 157 zeros following after it, and would look like the following number:

10,000,000,000,000,000,000,000,000,000,000,
000,000,000,000,000,000,000,000,000,000,000,
000,000,000,000,000,000,000,000,000,000,000,
000,000,000,000,000,000,000,000,000,000,000,
000,000,000,000,000,000,000,000,000

There has never been that number of people born on the face of the earth -ever- and will never be for eons of time to come. Yet Jesus was born about 2000 years ago and fulfilled all the requirements of those prophecies as spoken about Him in the Holy Bible. Dr. Stoner projected that if only eight prophecies were fulfilled, the statistical probability would be 10^{17}. Using a silver dollar for each of the 10^{17} numbers, silver dollars could cover the State of Texas two feet deep. Then if one of those silver dollars contained a special marking and one person was given just one chance to find the one specially marked silver dollar – that would be the one chance in 10^{17}.

A 10^{157} probability with silver dollars would be much the same, only with those two-foot deep silver dollars extending to cover the entire Unites States. The website for this information is found within the geocites.com website. Conduct a search on Science Speaks and Doctor Peter Stoner. These facts are also documented in the book *More Than A Carpenter* by Josh McDowell.

Jesus, however, accomplished all of those prophecies in His life span of just thirty-three years. Therefore, the statistical probability was actually increased to more than ***double*** that 10^{157} possibility that someone could fulfill all of those prophesies! So then, 'do the math' for yourself. Look at the statistical chances of Christ being who He said He was: the centuries-long prophesied Messiah. The facts of statistical math point to Christ Jesus being the Son of the one true living God. However, just doing the math and accepting the stats are not the guaranteed tickets for entry through Heaven's Pearly Gates. Asking forgiveness for all our wrongs and sins, and accepting Christ into our heart and life as Savior and Lord, and believing – **IS** the guaranteed ticket.

There is definitely a factual ingredient here with the math of 10^{157} for building faith in the fact of there being a God. God is a mathematical genius. He ran the numbers on His Son Christ for modern-day mankind to see. Yet, it is up to modern man for acceptance of those stats into their hearts. Then, just as the unseen wind can be felt – the unseen things of *factual statistics* can be revealed through the eyes of faith, for making the things of God a reality!

There is another interesting fact worth examining. This fact is found in the statistics of literary science. Ever thought about how the Holy Bible ranks in manuscript comparison with other writings of antiquity? In the field of literary science, part of the validation process for ancient manuscripts (MSS), is by the numbers of originally hand-scribed manuscripts identical to the earliest known original handwritten manuscript. Those manuscripts are not mimeographed, photographed, or photocopied in any form, but actual ancient hand-scribed copies.

The book of *Caesar* has ten copies from antiquity spanning about 1,000 years; it has an MSS of 10. Plato's *Tetralogies* has seven copies from antiquity spanning about 1,200 years; it has an MSS of 7. Tacitus' *Annals* has an MSS of 20 spanning 1,000 years. Pliny The Younger's *History* has an MSS of 7 spanning about 750 years. Suetonius' *DeVita Caesarum* has an MSS of 8 spanning about 800 years. Homer's *Iliad* has an MSS of 643 spanning about 500 years.

The MSS number and span of years vary with the various sources. In my illustration I wish to focus on Homer's *Iliad* for the literary world seems to hold that work has the most validated authentic piece of writing from antiquity. For Homer's *Iliad* I'll use an

MSS of 700, because in my readings, research results have revealed an MSS ranging from about 640 to just over 800. Therefore, an MSS of 700 would be a good average number for this illustration.

Continuing on, there are approximately 700 manuscripts of Homer's *Iliad,* which have been hand-scribed in antiquity, and are basically identical to the earliest known copy. Homer's *Iliad* has been widely accepted in the literary world as the oldest most validated authentic writing from antiquity. Thus, no one doubts the authenticity of Homer's original work.

Apply that very same literary scientific principle to the Old Testament of the Bible, and the results will come up to an MSS of approximately 8,000, over 1,600 years. Extend that concept to the New Testament and the results reveal an MSS of about 25,000, **AND**, its span of time only ranges about 50 years! Quite a significant difference!

Logic from these statistics alone would dictate that the Holy Bible is, in fact, the most validated authentic writing from antiquity that we have in the world. However, most people do not want to accept these statistical facts. Why? Consider the ole cliché, which says, 'I know what was said (or I know what I want), don't confuse me with the facts.' We just simply want things our own way, which does not include God's way most of the time.

An MSS of 700 spanning about 500 years, versus an MSS of 8,000 spanning 1,600 years, and an MSS of 25,000 spanning 50 years – is quite a significant validation factor for authentication! The MSS concept for the Bible is a very solid faith-building ingredient. It very factually illustrates the soundness for the uniquely high authenticity of the Holy Bible, **AND**, having faith in what it says! Sadly, though, too many people do not accept those sound facts. They prefer living lives of 'not being confused by the facts.'

In mentioning the preference of living 'not being confused by the facts,' why is it that God has to be proven, but everything else can be accepted seemingly Carte Blanc at face value? Why does the one true God, creator of the entire universe, have to prove His Divinity? Yet, all the other pagan gods of this world are accepted as is. Why does faith in God require empirical truths, proofs, and reasons? Yet, having faith in any other thing appears so easy, and even acceptable? Why does it seem that the God of all gods, must show, and prove, some Certification of Divinity? Yet, all the other lesser gods receive lavish credentials of supernatural origin, and most of those gods are just figments of imagination or carved icons.

It seems from the very beginning of mankind's existence, man has held high court on proving the existence of God. God has been on trial since mankind rebelled against Him. Just go out on a clear starry night, and look up into the heavens. Where did all those stars come from? Some entity, far more advanced than any of us, flung all of it into place long before this sphere of Earth came into existence. That entity is God. Quite simply, gazing into a telescope or a microscope for a while at the countless wonders those lenses can reveal, should settle the matter of God's existence.

This is the intersection where faith connects with reality; we need His guidance and we need to trust in Him. God always knows what is best, for He sees the future and we cannot. In order to arrive at the best destinations of life by excellently-lighted pathways, there must be something to make the connections. That something is called faith. Faith is invisible as the wind, it but touches the heart of God.

One last consideration on whether or not there is a God - the ***dedicated minister***. When one becomes a dedicated minister for God, not one who plays religion or uses it only as a vocation, that one will definitely find out the existence of evil. Satan, and all those on his side are forever assailing God's dedicated ministers. If there was a doubt in the mind about there being a God, becoming one of God's ***dedicated*** ministers will definitely erase that doubt.

If there is a devil for the evil,
then there must be a God for the good.

"The fool has said in his heart,
'There is no God...'"
(Psalm 14:1).

CHAPTER XXV
EPILOGUE

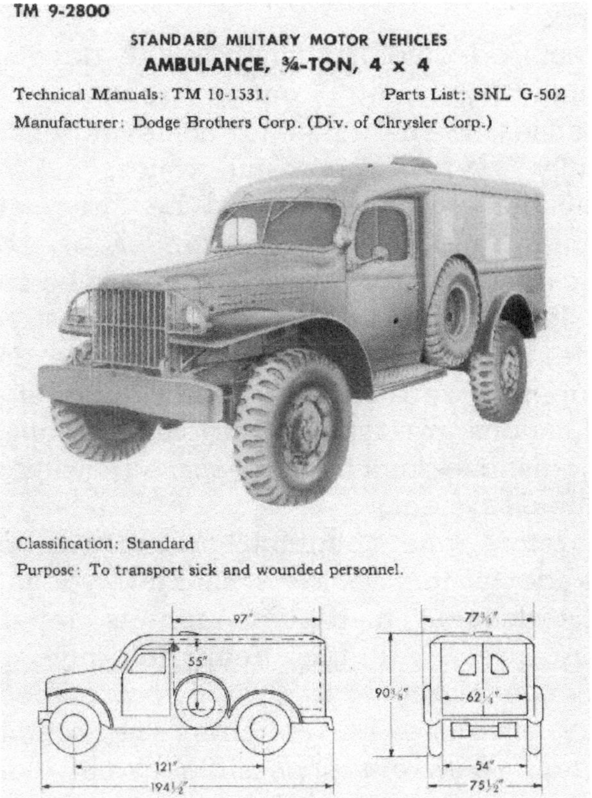
TM 9-2800

STANDARD MILITARY MOTOR VEHICLES

AMBULANCE, ¾-TON, 4 x 4

Technical Manuals: TM 10-1531. Parts List: SNL G-502

Manufacturer: Dodge Brothers Corp. (Div. of Chrysler Corp.)

Classification: Standard

Purpose: To transport sick and wounded personnel.

This book explored some very basic historical concepts for specifications as applicable for the military. Actual examples were provided along with some historical supporting data. In addition, some of my own personal experiences provided some examples. Without specifications, the world as we know it would not be as technical as it is, and would not be as enjoyable as it is either. Specifications provide a means for launching into the unknown realm – but with some *structured guidance*.

Structured guidance enables the developer to systematically map out carefully planned approaches, or strategies, for achieving end goals; hopefully very successful ones. Without specifications, there would be no seriously meaningful end result to any endeavor. *Without specifications*, nearly every venture would result in some level of disaster. Specifications actually provide invisible points of reference for the developer to zero in on a desired target result. Specifications are actually the *guiding light* for all the logistical functions encased within any operational system.

Therefore, what about the specifications for your life, reader of this book? What have you taken as the instructional map-guide for its repair and sustainment? All the years of my military experience taught me to always be prepared. The military instructs its members in contingency planning. With every possible event, there are possibilities of many unforeseen scenarios; and those scenarios must be adequately planned for. Thus, with contingency planning, direct and collateral damages can be minimized. What about applying the same concept of contingency planning toward *life in the hereafter*?

Contingency planning into the hereafter in worst-case scenarios is the *best exit strategy*. Contacting God in making advance preparations for leaving this life, when the time arrives – should be the ONLY *exit option* for anyone. What *specification* do you have for that *final exit option?*

Please do not wait until some serious situation crosses the pathway of your life, and you have nowhere to turn. Within the matrix of disastrous 'Oh My God' life situations, sound logical thinking and planning is very difficult. Additionally, very often during those events limited time is encountered for making things eternally right with our Maker – the Lord God. Why not settle that issue while reading this book, and before some personal or family cataclysmic event unravels the fabric of life? The process is simple and does not cost anything.

There is no transfer fee – it is free. Christ Jesus paid all the associated cost with His death on the cross about 2000 years ago.

However, if you chose in delaying that decision, do not forget about finding God while living; do find the time for cleaning up life's record through His Son, Christ Jesus, and inviting Him into your life. If you DO forget, and DO NOT find Him before this life is over, then is when you WILL certainly face the biggest **OH MY GOD** situation of your entire existence – and it will be ***forever*** too late! Use the *Finding God* section in the Addendum for inviting God into your life. The number of days allotted to each life is known only to God! Do not waste those allotted days!

Addendum I
Finding God

Finding God and accepting Him into your life in a legal sense is actually somewhat of a Pre-trial Hearing. Each one of us will stand before God some day and give an account of the things we have done in this life (reference Revelation 20:11-15). By asking God for forgiveness and accepting it before that time through His Son, Christ Jesus, the record book Heaven keeps on each of us is expunged of our transgressions. As we continue living in the Grace of God, ask forgiveness, and receive it, our record is continually expunged. Thus, we in effect, live out somewhat of a Pre-Trial Hearing until Judgment Day. Then, Christ, as our Attorney, will stand in for us.

Finding God and accepting His Son, Christ Jesus, is not complicated at all; it is actually very easy. It is as simple as '**ABC**.' The first of the ABCs is ***Admit***. Admit having sinned, being sorry for them, and being in need of God's forgiveness for those attitudes and actions not meeting God's standards. *"If we say that we have no sin, we deceive ourselves, and the truth is not in us. If we confess our sins, He is faithful and just to forgive us our sins and to cleanse us from all unrighteousness"* (1 John 1:8 and 9). *"For the wages of sin is death, but the gift of God is eternal life in Christ Jesus our Lord"* (Romans 6:23).

The second of the ABCs is ***Believe***. Believe that Christ Jesus died and rose again paying the penalty of death for our sins. God loves us so much, He provided a way of escaping that penalty. *"For God so loved the world that He gave His only begotten Son, that whoever believes in Him should not perish but have everlasting life"* (John 3:16). *"But as many as received Him, to them He gave the right to become children of God, to those who believe in His name"* (John 1:12).

The third of the ABCs is ***Confess***. Confess Christ Jesus as the Lord and Savior of your life. Confessing Jesus means yielding total control of our lives to Christ and telling someone else what He has done. If He is given ownership of our life, He becomes our new manager. *"...if you confess with your mouth the Lord Jesus and believe in your heart that God has raised Him from the dead, you will be saved...For whoever calls on the name of the Lord shall be saved"* (Romans 10:9 and 13).

A simple prayer for finding God can be as follows: 'Oh God, I have sinned and am sorry for them, and I am in need of You. Please forgive me. I believe Your Son, Christ Jesus, died and rose again for my sins. I confess Him now as Lord and Savior of my life. Thank you for saving me.' This simple prayer does not 'save' a person in and of itself. What is does, though, is provide the basic foundation for expressing one's faith to God in saving one from their transgressions, through the sacrifice Christ gave of Himself with His own death on the cross about 2000 years ago.

After saying that prayer, do two things preferably every day. First, read some in the Holy

Bible, and second, pray. Praying is just simply talking to God and is communication with Him. Prayer begins a relationship with God and nourishes it. Bible reading is further nourishment for the spiritual person reborn inside from the former person. It lays a solid foundation of knowledge and wisdom for building the personal relationship with God.

Then, find a good Bible teaching church to attend. Christians need other Christians around them, praying for one another, and gathering themselves together worshipping God. Regular church attendance further nourishes the new life begun with God. Churches are like people in some respects – they have personalities. Find a church where your life can blend in and where the Spirit of God is freely flowing.

Do not wait to try and find God toward the end of life. Those are usually the hardest times to pray. We are never guaranteed how much remaining time for closure there will be...

ADDENDUM II
FAITH PASSAGES

A few passages from the Holy Bible for guidance through life's pathways:

1. Psalm 9:12 –*...He does not forget the cry of the humble.*
2. *As far as the east is from the west, so far has He removed our transgressions from us. As a father pities his children, so the LORD pities them that fear Him. For He knows our frame; He remembers that we are dust* (Psalm 103:12, 13 and 14).
3. Psalm 119:105 – *Your word is a lamp to my feet and a light to my path.*
4. Psalm 139:17 and 18 – *How precious also are Your thoughts to me, O God! How great is the sum of them! If I should count them, they would be more in number than the sand; when I awake, I am still with You.*
5. Isaiah 1:18 – "*'Come now, and let us reason together,' says the LORD, though your sins are like scarlet, they shall be as white as snow; though they are red like crimson, they shall be as wool*'".
6. Isaiah 53:6 – *All we like sheep have gone astray; we have turned, every one, to his own way; and the LORD has laid on Him the iniquity of us all.*
7. Lamentations 3:22 and 23 – *Through the LORD's mercies we are not consumed, because His compassions fail not. They are new every morning: great is Your faithfulness.*

8. Ezekiel 18:32 – "'*For I have no pleasure in the death of one who dies,' says the* LORD GOD. *'Therefore turn and live!'*"
9. Matthew 7:9-11 – *...what man is there among you who, if his son asks for bread, will he give him a stone? Or if he asks for a fish, will he give him a serpent? If you then, being evil* [of carnal nature], *know how to give good gifts to your children, how much more will your Father who is in heaven give good things to those who ask Him!*
10. Matthew 10:29 and 31 – *Are not two sparrows sold for a copper coin? And not one of them falls to the ground apart from your Father's will...you are of more value than many sparrows.*
11. Matthew 11:28 – *Come to Me, all you who labor and are heavy laden, and I will give you rest.*
12. John 3:16 – *For God so loved the world that He gave His only begotten Son, that whoever believes in Him should not perish but have everlasting life.*
13. John 6:37 – *...the one who comes to Me I will by no means cast out.*
14. John 10:10 – *...I have come that they may have life, and that they may have it more abundantly."*
15. Romans 3:23 – *For all have sinned and fall short of the glory of God.*
16. Romans 5:8 – *But God demonstrates His own love toward us, in that while we were still sinners, Christ died for us.*
17. I John 1:9 – *If we confess our sins, He is faithful and just to forgive us our sins and to cleanse us from all unrighteousness*
18. I John 2:1 – *And if anyone sins, we have an Advocate with the Father, Jesus Christ the righteous.*

ADDENDUM III
QUOTES

1. Leadership is a potent combination of strategy and character. But if you must be without one, be without the strategy (General Norman Schwarzkopf).
2. We need to learn to set our course by the stars, not by the lights of every passing ship (General Omar Bradley)
3. Lead me, follow me, or get the hell out of my way (General George Patton).
4. The soldier is the Army. No army is better than its soldiers. The Soldier is also a citizen. In fact, the highest obligation and privilege of citizenship is that of bearing arms for one's country (General George Patton).
5. The military don't start wars. Politicians start wars (General William Westmoreland).
6. Whoever said the pen is mightier than the sword obviously never encountered automatic weapons (General Douglas MacArthur).
7. I can imagine no more rewarding a career. And any man who may be asked in this century what he did to make his life worthwhile, I think can respond with a good deal of pride and satisfaction: 'I served in the United States Navy (Lieutenant Junior Grade and U.S. President John F. Kennedy).
8. Ask not what your country can do for you, but want you can do for your country (President John F. Kennedy).

9. Don't start vast projects with half vast ideas (Unknown).
10. Never tell people how to do things. Tell them what to do and they will surprise you with their ingenuity (General George Patton).
11. If they want peace, nations should avoid the pin-pricks that precede cannon shots (Napoleon Bonaparte).
12. Imagination rules the world (Napoleon Bonaparte).
13. Impossible is a word to be found only in the dictionary of fools (Napoleon Bonaparte).
14. A soldier will fight long and hard for a bit of colored ribbon (Napoleon Bonaparte).
15. A picture is worth a thousand words (Napoleon Bonaparte).
16. Discipline is the soul of an army. It makes small numbers formidable; procures success to the weak, and esteem to all (General and U.S. President George Washington).
17. No good decision was ever made in a swivel chair (General George Patton).
18. The best morale exists when you never hear the word mentioned. When you hear it it's usually lousy (General and U.S. President Dwight D. Eisenhower).
19. Live for something rather than die for nothing (General George Patton).
20. I have nothing to offer but blood, toil, tears, and sweat (Sir Winston Churchill).
21. Experts often possess more data than judgment (General Colin Powell).

"For what profit is it to a man
if he gains the whole world,
and loses his own soul?

Or what will a man give in
exchange for his soul?"

~ Matthew 16:26 ~

This book was produced and
published in association with:

Dream Weaver Ministries, LLC
Springfield, Ohio, USA
www.dreamweaverministries.com

*"Keep your heart with all diligence,
for out of it spring the issues of life"*
(Proverbs 4:23)

Made in the USA
Middletown, DE
04 November 2026

36535578R00136